JN175825

日経サイエンスで鍛える科学英語

ノーベル賞科学者編

Scientific English with
SCIENTIFIC AMERICAN®/
NIKKEI SCIENCE

日経サイエンス編集部［編］

発行 日経サイエンス社

まえがき

本書は月刊科学雑誌『日経サイエンス』に翻訳掲載した *SCIENTIFIC AMERICAN* 誌の記事から一部を抜粋し，原文と対照して読めるようにしたものであり，『日経サイエンスで鍛える科学英語』（2011年刊）および『日経サイエンスで鍛える科学英語2［読解編］』（2013年刊）の続編にあたる。

「興味深い科学ストーリーを英語で読むことによって，楽しみながら科学英語に親しむ」という一石二鳥の狙いはこれまでの2冊と同じであるが，本書では特にノーベル賞科学者が執筆した記事を集めた。最も古い記事はマリー・キュリーによる1908年の「電気と物質の最新理論」，最新はアハメッド・ズウェイルによる2010年の「極微の世界をとらえるナノムービー」であり，全体を通読することによって過去1世紀の科学の進展を概観できる内容とした。また，本書編纂中に2015年のノーベル物理学賞が梶田隆章・東京大学教授に贈られるという喜ばしいニュースが飛び込んできたため，梶田氏が共著したニュートリノに関する解説記事をやや長目に収録し，読者の関心に応えることとした。マリー・キュリーから梶田博士まで，科学の歩みに思いをめぐらせていただきたい。

こうした構成が可能になったのは，*SCIENTIFIC AMERICAN* の創刊が今からちょうど170年前の1845年にさかのぼり，米国で継続して発行されている雑誌のなかで最も長い歴史を誇ることによる。一般向け科学雑誌として，その時々の最先端の科学トピックが研究者自身あるいはサイエンスライターによる明快な英語で綴られてきた。その質の高さには定評があり，著者が後にノーベル賞を受賞する例が少なくない。本書に収録したのはそうした記事の一部である。

SCIENTIFIC AMERICAN の日本語版は SCIENTIFIC AMERICAN 社と日本経済新聞社の協力の下に1971年に『サイエンス』として創刊した（1990年10月号から現在の題号『日経サイエンス』に）。それ以前の記事については日本語訳が出版されていなかったわけだが，本書は *SCIENTIFIC AMERICAN* が2011年から2014年にかけて4回編纂したノーベル賞関連特集をベースとしている。ノーベル賞科学者が執筆した過去記事からエッセンスを抜粋したもので，日経サイエンス誌上では2011年から2014年までの各11月号に掲載した。これらから既刊の『日経サイエンスで鍛える科学英語2』に収録したものを除くなど

の調整をしたうえ，ウェブのみに掲載された一部記事（ウィリアム・ブラッグ著「物質の原子構造を探るX線の指」）および梶田博士によるニュートリノの記事を加えて構成した。

ノーベル賞の分類に準じて，「生理学・医学賞」「物理学賞」「化学賞」の3章に分けた。各章ごとに，記事は執筆・掲載時期の古い順に並べてある。順に読み進むと，科学的思考と研究の変遷を把握できるだろう。歴史的な記事も多く，現代科学の最先端に焦点を当てたものでは必ずしもないが，ノーベル賞創設（1901年）以来100年あまりの科学研究の進展が現在につながっていることを読み取れるはずである。ただ，「ノーベル賞受賞者本人が *SCIENTIFIC AMERICAN* に執筆した記事」だけで20世紀の科学史を系統的にカバーするのはもとより不可能であり，その点はご容赦いただきたい。

英文の右側に語注「Vocabulary」を併記したほか，主な科学技術用語について簡単な説明「Technical Terms」を掲載した。理解に必要と思われる基本的な科学用語を中心に解説したが，歴史的記事には現在ではすでに耳慣れなくなった装置名や用語が登場する場合も見られたので，それらを説明した例もある。

一般に本誌「日経サイエンス」の訳文は，読みやすさを重視して，必ずしも逐語訳にはしていない。また，科学的な理解を助けるために，原文にない内容を補足したり，部分的に順序を入れ替えたりしている例がある。これをそのまま原文と対照して読むと齟齬が生じるため，本書では特に違和感が強くなると思われる部分については原文に即した訳文に改めた。とはいえ，英語の試験で求められるような逐語訳ではないことをお断りしておく。

なお，本誌ウェブサイト（http://www.nikkei-science.com）には「英語で読む日経サイエンス」というコンテンツがある。各号に掲載した主要記事から毎月1～2本を選び，その冒頭から数パラグラフの原文・訳文を並べて表示したものだ。2005年夏にスタートし，2015年11月時点で180本あまりをアップしてある。また誌上では2009年5月号から「今月の科学英語」を連載している。これはその号に登場した科学用語をいくつか選んで原文とともに掲載し，簡単な解説を加えたものだ。既刊書2冊を含め，これらも併せてご利用いただきたい。

2015年12月

日経サイエンス編集部

日経サイエンスで鍛える
科学英語
ノーベル賞科学者編

目次

2 物理学賞

素粒子から
宇宙まで

3　化学賞
物質と生命の
振る舞い

凡例　※翻訳記事では，よりわかりやすく読みやすい文章にするため，内容を一部補強するなどの変更をしている場合があり，必ずしも英文記事の逐語訳とはなっていない。また，改行位置の変更など日経サイエンスの誌面掲載時とは異なる文章になっている場合がある。
※各記事の初出と抜粋記事掲載の時期を，表題ページの下の部分に，SCIENTIFIC AMERICANおよび日経サイエンス（一部は旧題号『サイエンス』）の掲載号で示した。

装丁：八十島博明
カバーイラスト：中村知史
DTP：GRID

日経サイエンスで鍛える
科学英語

ノーベル賞科学者編

生理学・医学賞
命の神秘に迫る

ウイルス Viruses

神経インパルス The Nerve Impulse

生命の起源 The Origin of Life

遺伝物質の構造 The Structure of the Hereditary Material

動物の求愛行動 The Courtship of Animals

皮膚移植 Skin Transplants

行動の進化 The Evolution of Behavior

神経細胞と行動 Nerve Cells and Behavior

免疫系の機能 The Immune System

視覚の脳内機構 Brain Mechanisms of Vision

ニューロンレベルでみた学習 The Biological Basis of Learning and Individuality

プリオン病はどこまで解明されたか The Prion Diseases

匂いの分子生物学 The Molecular Logic of Smell

テロメアとがん Telomeres, Telomerase and Cancer

真核細胞はどのように生まれたか The Birth of Complex Cells

Viruses

ウイルス（1951年掲載）

F. M. バーネット（1960年受賞）

フランク・バーネット（Frank M. Burnet, 1899～1985年）は免疫の研究で知られるオーストラリアのウイルス学者。細菌に感染するウイルスであるバクテリオファージをはじめ，ウイルスが複製する仕組みや免疫系との関連に関して多くの事実を発見した。1960年,「後天的免疫寛容の発見」で英国の生物学者ピーター・メダワー（32ページ参照）とともにノーベル生理学・医学賞を受賞した。

バーネットが免疫の研究に主軸を移したのは1950年代だが，ウイルスについて解説した1951年のこの記事でも免疫の視点からの記述が強調されている。SCIENTIFIC AMERICANの記事掲載はノーベル賞を受賞する9年前だ。

初出：SCIENTIFIC AMERICAN May 1951
抜粋掲載：SCIENTIFIC AMERICAN June 2011, 日経サイエンス 2011年11月号

A virus can be defined as a microorganism, considerably smaller than most bacteria, which is capable of multiplication only within the living cells of a susceptible host. The practical control of a virus disease nearly always depends essentially on obtaining an understanding of the means by which the balance between the virus and the host is maintained in nature and how it can be modified in either direction by biological accident or by human design. In the approach to such an understanding two important related concepts have emerged—"subclinical infection" and "immunization."

A subclinical infection is one in which the infected person gives no sign of any ill effect. In a population attacked by an infectious disease, subclinical infections often greatly outnumber those severe enough to produce unmistakable symptoms of the disease. For example, when a child comes down with a paralyzing attack of poliomyelitis, a careful examination of the rest of the family will commonly reveal that all the other children have the virus in their intestines over a period of a week or two, but they either show no symptoms at all or have only a mild, nondescript illness. Fortunately even a subclinical infection produces heightened resistance or immunity to the virus for a period after the attack. This capacity of mild or subclinical infection to confer immunity is probably the greatest factor in maintaining a tolerable equilibrium between man and the common virus diseases. The trouble is that viruses are labile beings, liable to undergo mutation in various directions, and a virus that causes only mild infection may evolve into one far more deadly.

Vocabulary

bacteria 細菌
▶ Technical Terms
multiplication 増殖
susceptible 影響を受けやすい, 病原体に感染しやすい
host 宿主
control 抑制
modify 変える, 変更する

subclinical infection 不顕性感染, 無症状感染
immunization 免疫化, 予防接種

sign 病気の徴候, 病状
population 集団
infectious disease 感染症
unmistakable 明白な
symptoms 症状
comes down with~ (〜の)病気になる
poliomyelitis ポリオ, 小児麻痺
intestine 消化管

nondescript 分類しがたい
immunity 免疫
▶ Technical Terms

confer 付与する
equilibrium 平衡状態

labile 変わりやすい
liable 〜しやすい
mutation 変異
▶ Technical Terms

Technical Terms

細菌(**bacteria**)　核膜で覆われた明確な細胞核を持たない原核生物。生物の分類に関する現在の考え方では, 生物を細菌と古細菌(**archaea**), 真核生物の3つのドメインに大別する。古細菌と区別するために「真正細菌」ともいう。

免疫(**immunity**)　動物が外来や内因性の異物を認識して体内から排除する生理的な仕組み。これを細胞・分子レベルで解明する研究は20世紀に大きく進展した。詳細は未知の部分も多く, 現在も活発な研究が続けられている。

変異(**mutation**)　遺伝子の突然変異などによって生じる形質の相違。

One cannot claim that there is full agreement about the nature of immunity to viruses, but it is possible to offer a simplified account which most virologists would accept. This interpretation is that all immunity to viruses is mediated through antibody. Antibodies can be described as modified blood-protein molecules capable of attaching themselves firmly to the specific virus or other invading organism that provoked their production by the body. If a sufficient number of antibody molecules can attach themselves to a virus particle, they have a blanketing effect which prevents the virus' attachment to the host cell and its multiplication within the cell. Antibody appears in the blood a few days after infection and reaches a peak in two to three weeks. The body continues to produce antibody at a slowly diminishing level long after recovery—in some diseases, such as measles and yellow fever, for the whole of life.

Vocabulary

account 説明
interpretation 解釈，説明
mediate 仲介する，結果を左右する
antibody 抗体
▶ Technical Terms
protein タンパク質
provoke 引き起こす
blanket 妨害する，何かを覆って妨げる
multiplication 増殖，複製
measles はしか，麻疹
yellow fever 黄熱病

Technical Terms

抗体(**antibody**) 免疫応答を引き起こす物質を抗原(**antigen**)といい，この刺激によって抗原と特異的に結合するタンパク質が生体内に作られる。これらのタンパク質は抗体と総称され，免疫における防御反応の重要な一翼を担っている。

ウイルスは一種の微生物といえるが，ほとんどの細菌よりもかなり小さく，自身が感染した宿主細胞のなかでのみ複製・増殖できる。ウイルス病を抑える方法はほとんど常に，ウイルスと宿主の間のバランスが自然界で保たれている方法を理解すること，そして生物学的アクシデントあるいは人間の目論見によってそのバランスをいずれかの方向にどうすれば変えられるかを理解することにかかっている。そうした研究のなかで，相互に関連した2つの重要な概念が生まれた。「不顕性（無症状）感染」と「免疫化（予防接種）」だ。

不顕性感染は感染者が何の病状も示さないタイプの感染だ。ある集団が感染症に攻撃されたとき，明らかな症状を示す重症者よりも不顕性感染の例のほうがずっと多いことがしばしばだ。例えばある子どもがポリオに倒れたとき，その家族を注意深く調べると，他の子どもたちがみな1～2週間にわたって消化管内にポリオウイルスを抱えているにもかかわらず，まったく症状がないか，ごく軽くて診断がつかないことがよくある。幸いなことに，不顕性感染であっても，感染後しばらくは病原ウイルスに対する抵抗，つまり免疫が強まる。軽い感染や不顕性感染が免疫を与えるこの能力は，人間とよくあるウイルス病が許容可能な平衡状態を維持しているおそらく最大の要因だろう。問題は，ウイルスが変わりやすい存在であり，さまざまな方向に変異する傾向があること，そしてごく軽い感染症を起こすだけだったウイルスがずっと致死的なものに進化する可能性があることだ。

ウイルスに対する免疫の本質について見方が完全に一致しているとはいえないが，大半のウイルス学者が受け入れると思われる簡単な説明を提供することはできる。ウイルスに対する免疫はすべて，抗体によって仲介されているとする解釈だ。抗体は血液タンパク質分子が変化して，その抗体の生産をそもそも身体に引き起こしたウイルスなど，特定の侵入生物にしっかり結合する能力を持つようになったものだといえる。もし十分な数の抗体分子がウイルス粒子に結合すれば，ウイルス粒子が宿主細胞にくっつくのを防ぐ効果を発揮し，細胞内でのウイルス増殖を妨げることができる。抗体は感染から数日後に血液中に現れ，2～3週間のうちにピークに達する。その後も身体は徐々に量を減らしながら，回復後も長期にわたって抗体を作り続ける。麻疹や黄熱病など一部の病気では，これが生涯にわたって続く。

The Nerve Impulse
神経インパルス（1952 年掲載）

B. カッツ（1970 年受賞）

ベルンハルト・カッツ（Bernhard Katz，1911 ～ 2003 年）はドイツ出身の英国の生理学者。1970 年,「神経末梢部における液性伝達物質，およびその貯蔵・解離・不活化の機構に関する発見」によって，米国の生化学者ジュリアス・アクセルロッド（Julius Axelrod，1912 ～ 2004 年）およびスウェーデンの生理学者・薬理学者ウルフ・フォン・オイラー（Ulf von Euler，1905 ～ 1983 年）とともにノーベル生理学・医学賞を受賞した。

この記事は神経細胞（ニューロン）が電気信号をどのように生み出し，それが神経線維に沿っていかに伝達されていくのか，細胞レベルのメカニズムについて述べている。細胞膜を通してニューロンにイオンが流入することで膜電位が変化する機構は神経インパルス伝達の基本だ。ノーベル賞の授賞理由はニューロンの接合点であるシナプスにおける神経伝達物質に関するものだが，カッツはそれにつながる基礎を築いた。

SCIENTIFIC AMERICAN がこの記事を掲載したのはカッツがノーベル賞を受賞するよりも 20 年近く前。ここに述べられている 1952 年当時の最先端の知見はいまから見ると教科書的な常識に思えるものの，そうした知見の積み重ねが科学を駆動している。

初出：SCIENTIFIC AMERICAN November 1952
抜粋掲載：SCIENTIFIC AMERICAN June 2011, 日経サイエンス 2011 年 11 月号

Some of the foremost nerve physiologists have considered it worthwhile to study and analyze the properties of nerve fibers from the point of view of the cable engineer. The nerve fiber is in effect a chain of relay stations—a device with which the communications engineer is thoroughly familiar. Each point along the fiber receives an electric signal from the preceding point, boosts it to full strength and so enables it to travel a little farther. It is a peculiar combination of a cable (of very defective properties) with an automatic relay mechanism distributed all along the transmission line. Before the electric signal has had a chance to lose its strength, it stimulates the fiber, releases local energy resources and is renewed. The electric potential difference across one point of the fiber membrane serves to excite the region ahead, with the result that this region now contributes, at its own expense, a greatly amplified electric signal, capable of spreading to and exciting the next region. Experiments have fully confirmed this concept of how a nerve fiber transmits a signal.

When a current passes through the membrane, partially discharging the membrane surface and thus reducing the electric field, this makes the membrane more permeable to sodium. Positive sodium ions begin to flow inward and further reduce the negative charge on the inside. Thus, the electric field across the membrane is further reduced, the sodium permeability continues to rise, more sodium enters, and we have the elements of a self-reinforcing chain reaction. The flow of sodium into the fiber continues until the fiber interior has been charged up to such a high positive level that sodium ions are electrostatically repelled. Now we can understand the basis of the all-or-none reaction of nerve cells: they generate no current until the "ignition point" is approached. Once this point is passed, the production of "sodium current" proceeds toward saturation and runs through a cycle of its own, no longer under the control of the original stimulus.

Vocabulary

foremost 一流の
nerve fiber 神経線維
in effect 実質的に
relay station 中継基地

defective 欠陥のある

stimulate 刺激する

electric potential 電気的ポテンシャル。ここでは膜電位
membrane 細胞膜

discharge 放電する

permeable 透過性の
sodium ナトリウム

permeability 透過性

self-reinforcing 自己増幅の
chain reaction 連鎖反応

流の神経生理学者の幾人かは，神経線維の特性を電気通信技術の視点から調査・解析することに十分な価値があると考えてきた。神経線維は実質的に，中継基地がつながった鎖であって，これは電気通信技術者ならすっかりお馴染みの装置だ。神経線維沿いの各中継点が前の点から電気信号を受け取り，最大限それを強め，もう少し遠くまで伝わるようにする。電線（ただし非常に不完全な特性の）と自動的な中継機構という独特の組み合わせが，伝送路すべてにわたって分散しているのだ。この電気信号は強さを失う前に神経線維を刺激し，その場のエネルギー源からエネルギーを引き出して強さを取り戻す。神経線維上のある一点での膜電位（細胞膜表面と内部の電位差）は，その前方の領域を興奮させるように作用し，その結果，自分自身を犠牲にしながら電気信号を大きく増幅するのに寄与して，信号が次の領域に伝わってそこを刺激できるようにしている。神経線維がこうして信号を伝えているという考えは，実験を通じて完全に確かめられている。

神経細胞の細胞膜を電流が通過していくとき，細胞膜の表面を部分的に放電させて電場を弱めるため，そこの膜はナトリウムイオンを通しやすくなる。プラスの電荷を持つナトリウムイオンが細胞内に流入し始め，内部の負電荷をさらに減らす。こうして細胞膜を挟む電場はさらに弱まり，ナトリウムの透過性は上がり続け，より多くのナトリウムが流入し，という具合に，自己増幅型の連鎖反応が進む要素がそろう。ナトリウムの流入は，神経線維内部の電位が十分に高まってナトリウムイオンがそれ以上入ろうとしても静電気力によって排斥されるまで続く。これによって神経細胞の“全か無か”の反応の基礎がわかった。神経細胞は「発火点」に近づくまでは電流を生じない。そしていったんこのポイントを超えると，“ナトリウム流”が生じて飽和点に向かって進み，最初の刺激によるコントロールをもはや受けずに自分自身の反応サイクルを進めていく。

The Origin of Life

生命の起源（1954年掲載）

G. ウォルド（1967年受賞）

ジョージ・ウォルド（George Wald, 1906～1997年）は網膜の色素の研究で知られる米国の科学者。1967年,「視覚の化学的・生理学的基礎過程に関する発見」によって，スウェーデンのラグナー・グラニト（Ragnar Granit, 1900～1991年）および米国の生理学者ハルダン・ケファー・ハートライン（Haldan Keffer Hartline, 1903～1983年）とともにノーベル生理学・医学賞を受賞した。

ウォルドは1950年代，網膜から抽出した色素が吸収する波長を解析し，目の色覚に迫る基礎を築いた。こうした研究がノーベル賞の授賞理由となっているが，以下の記事はその内容を述べたものではない。「そもそも生命がいかに生まれたのか」という大きな謎について，1950年代当時の知見に基づいて考察したエッセイだ。いまから半世紀前，新たな生化学的手法によってこの謎に迫ろうとする科学者の思いを伝えている。

抜粋記事で述べられているノーベル化学賞受賞者ユーリーの実験は記事掲載の前年，1953年に行われた。「ユーリー・ミラーの実験」として有名で，著者がこの画期的実験に大いに刺激されたのは疑いない。ちなみにユーリーに対するノーベル化学賞（1934年）の授賞理由は「重水素の発見」。

初出：SCIENTIFIC AMERICAN August 1954
抜粋掲載：SCIENTIFIC AMERICAN June 2011, 日経サイエンス 2011年11月号

Organic molecules form a large and formidable array, endless in variety and of the most bewildering complexity. To understand how organisms originated we must first of all explain how such complicated molecules could come into being. To make an organism requires not only a tremendous variety of these substances, in adequate amounts and proper proportions, but also just the right arrangement of them. Structure here is as important as composition—and what a complication of structure! The most complex machine man has devised—say, an electronic brain—is child's play compared with the simplest of living organisms.

Recently Harold Urey, Nobel laureate in chemistry, has become interested in the degree to which electrical discharges in the upper atmosphere may promote the formation of organic compounds. One of his students, S. L. Miller, performed the simple experiment of circulating a mixture of water vapor, methane (CH_4), ammonia (NH_3) and hydrogen—all gases believed to have been present in the early atmosphere of the earth—continuously for a week over an electric spark. The circulation was maintained by boiling the water in one limb of the apparatus and condensing it in the other. At the end of the week the water was analyzed by the delicate method of paper chromatography. It was found to have acquired a mixture of amino acids! Glycine and alanine, the simplest amino acids and the most prevalent in proteins, were definitely identified in the solution, and there were indications it contained aspartic acid and two others. The yield was surprisingly

Vocabulary

organic molecule 有機分子
formidable 恐るべき，膨大な
bewildering 途方に暮れるほどの，あきれるほどの

proportion 比率
arrangement 配置
composition 組成

child's play 簡単なこと，たかが知れたこと

Nobel laureate ノーベル賞受賞者。laureate は栄冠を受けた人
discharge 放電

limb 周縁部，へり
condense 凝縮（液化）させる
paper chromatography ペーパークロマトグラフ法
▶ Technical Terms
amino acid アミノ酸
▶ Technical Terms
glycine グリシン
alanine アラニン
indication 兆候，しるし
aspartic acid アスパラギン酸
yield 収量

Technical Terms

ペーパークロマトグラフ法（**paper chromatography**） クロマトグラフィーは物質が持つ電荷や吸着力，質量などの違いを利用して物質を種類別に分ける技法のこと。ペーパークロマトグラフは濾紙を用いる簡単なもので，現代ではもっと高度な技術がいろいろある。

アミノ酸（**amino acid**） 化学的にはアミノ基とカルボキシル基を持つ有機化合物のこと。タンパク質は種類の異なるアミノ酸がたくさんつながったもので，アミノ酸の構成とその順序によって異なる形と機能を持つタンパク質になる。生物のタンパク質を構成するアミノ酸は本文中に登場するグリシンやアラニンなど 20 種類ある。

high. The amazing result changes at a stroke our ideas of the probability of spontaneous formation of amino acids.

Recently several particularly striking examples have been reported of the spontaneous production of familiar types of biological structure by protein molecules. Cartilage and muscle offer some of the most intricate and regular patterns of structure to be found in organisms. A fiber from either tissue presents under the electron microscope a beautiful pattern of cross striations of various widths and densities, very regularly spaced. The proteins that form these structures can be coaxed into free solutions and stirred into a completely random orientation. Yet on precipitating, under proper conditions, the molecules realign with regard to one another to regenerate with extraordinary fidelity the original patterns of the tissues.

We have therefore a genuine basis for the view that the molecules of our oceanic broth will not only come together spontaneously to form aggregates but in doing so will spontaneously achieve various types and degrees of order.

Vocabulary

at a stroke 一気に，一撃で
spontaneous 自発的な，自然に起こる

cartilage 軟骨
intricate 複雑な，入り組んだ

electron microscope 電子顕微鏡
cross striations 横紋
▶ Technical Terms
coax なんとかして〜にする

precipitate 沈殿する
realign 再編成する
regenerate 再生する，再び作り出す

genuine 本物の，正真正銘の
broth 煮出し汁，培養液

aggregate 集合体

Technical Terms

横紋（**cross striations**）　一般に横方向に伸びる縞模様のことで，生体組織では筋細胞に見られるものがよく知られる。横紋がある筋繊維が「横紋筋」。

有機化合物は実に多種多様であり，極めて複雑だ。生物がどのように出現したかを知るには，まずこれらの複雑な分子がどのように生じえたのかを説明しなければならない。生物体が生まれるには，これらの非常に多様な物質が適切な量と比率で必要になるだけでなく，正しく配置しなければならない。そうした構造が組成と同じくらい重要だが，この構造がまた何と複雑なことか！　人間が考案した最も複雑な機械，例えば電気式のコンピューターでさえ，最も単純な生物体と比べてもたかが知れている。

ノーベル化学賞受賞者のユーリー（Harold Urey）は最近，上層大気中の放電が有機化合物の生成をどれほど促進するか興味を抱いた。彼の指導学生ミラー（S. L. Miller）は，水蒸気とメタン（CH_4），アンモニア（NH_3），水素の混合物を１週間循環させながら電気火花を加える単純な実験を行った。これらのガスはどれも地球初期の大気に存在していたと考えられている物質だ。装置の一端で水を沸騰させ他端で凝縮することで，ガスの循環を維持した。１週間後，巧妙なペーパークロマトグラフ法によってこの水を分析した。そして複数のアミノ酸が発見された！　グリシンとアラニン，つまり最も単純なアミノ酸でタンパク質に最も広く含まれている分子が溶液中に確かに特定されたほか，アスパラギン酸など３種のアミノ酸の兆候も見つかった。収量は驚くほど高い。この驚異的な結果によって，アミノ酸が自発的に生成する可能性に関する私たちの考え方は一変した。

最近，よく知られた生体構造がタンパク質分子から自発的にできあがるという実に驚くべき例がいくつか報告された。軟骨と筋肉は生物体に見られる最も複雑にして規則的なパターンだ。どちらの組織も，そこに含まれている繊維を電子顕微鏡で観察すると，さまざまな幅と密度，極めて規則的な間隔を持った横紋の美しいパターンが見られる。これらの構造を作り上げているタンパク質を溶液中に遊離し，それをかき混ぜて完全にバラバラな方向を向かせることができる。しかし，それが沈殿する過程で，適切な条件では，分子が互いに再配列して元の組織に見られたパターンを驚くべき忠実度で再現する。

したがって，原始の海に含まれていた分子が自発的に集まって集合体を作るだけでなく，それによってさまざまな類型と秩序を自然に生み出すという見方に，確かな根拠が得られた。

The Structure of the Hereditary Material

遺伝物質の構造（1954 年掲載）

F. H. C. クリック（1962 年受賞）

フランシス・ハリー・コンプトン・クリック（Francis Harry Compton Crick, 1916 ～ 2004 年）は DNA の二重螺旋構造を発見した英国の科学者。1962 年，この業績によって米国の分子生物学者ジェームズ・ワトソン（James D. Watson, 1928 年～）および英国の生物物理学者モーリス・ウィルキンス（Maurice H. F. Wilkins，1916 ～ 2004 年）とともにノーベル生理学・医学賞を受賞した。

この画期的発見が *Nature* 誌に発表されたのは 1953 年で，クリックはそれから 1 年足らずのうちに以下の一般向け解説記事を SCIENTIFIC AMERICAN に執筆している。もちろんノーベル賞が授与される前だ。

クリックは物理学者としてスタートしたが，第二次世界大戦後に生物学に転向した。ノーベル賞受賞後は米国のソーク研究所などで研究生活を送り，特に 1990 年代以降は「なぜ脳から意識が生まれるのか」という問題に取り組んだ。このテーマについて述べた記事を共同研究者のクリストフ・コッホ（Christof Koch）との共著で SCIENTIFIC AMERICAN に執筆している（「意識とは何か」，日経サイエンス 1992 年 11 月号）。

初出：SCIENTIFIC AMERICAN October 1954
抜粋掲載：SCIENTIFIC AMERICAN June 2011, 日経サイエンス 2011 年 11 月号

It is now known that DNA consists of a very long chain made up of alternate sugar and phosphate groups. The sugar is always desoxyribose. While the phosphate-sugar chain is perfectly regular, the molecule as a whole is not, because each sugar has a "base" attached to it. Four different types of base are commonly found: two of them are purines, called adenine and guanine, and two are pyrimidines, known as thymine and cytosine. So far as is known the order in which they follow one another along the chain is irregular, and probably varies from one piece of DNA to another. Although we know from the chemical formula of DNA that it is a chain, this does not in itself tell us the shape of the molecule, for the chain, having many single bonds around which it may rotate, might coil up in all sorts of shapes.

J. D. Watson and I, working in the Cavendish Laboratory at Cambridge, were convinced that we could get somewhere near the DNA structure by building scale models based on the x-ray patterns obtained by M.H.F Wilkins, Rosalind Franklin and their co-workers at King's College London. To get anywhere at all we had to make some assumptions. The most important one had to do with the fact that the crystallographic repeat did not coincide with the repetition of chemical units in the chain but came at much longer intervals. A possible explanation was that all the links in the chain were the same but the x-rays were seeing every tenth link, say, from the same angle and the others from different angles. What sort of chain might produce this pattern? The answer was easy:

Vocabulary

sugar 糖
phosphate group リン酸基
desoxyribose デオキシリボース。deoxyribose に同じ
base 塩基

purine プリン塩基
▶ Technical Terms
adenine アデニン
guanine グアニン
pyrimidine ピリミジン塩基
▶ Technical Terms
thymine チミン
cytosine シトシン
chemical formula 化学式
single bond 単結合

scale model スケールモデル

assumption 仮定，想定
crystallographic 結晶学的な
coincide with ～と一致する
repetition 反復

Technical Terms

プリン塩基(**purine base**)　プリンそのものは分子式 $C_5H_4N_4$ の芳香族化合物で，プリン環と呼ばれるこの構造を基本骨格とする塩基がプリン塩基。DNA の 4 種類の塩基のなかではアデニンとグアニンがこの仲間に属する。本文では単に "purine" と述べているが，これは "purine base" の意味だ。

ピリミジン塩基(**pyrimidine base**)　ピリミジン $C_4H_4N_2$ も芳香族化合物の一種で，これを骨格とする塩基がピリミジン塩基。DNA の 4 種類の塩基のなかではチミンとシトシンがこの仲間に属する。

the chain might be coiled in a helix. The distance between crystallographic repeats would then correspond to the distance in the chain between one turn of the helix and the next.

This particular model contains a pair of DNA chains wound around a common axis. The two chains are linked together by their bases. A base on one chain is joined by very weak bonds to a base at the same level on the other chain, and all the bases are paired off in this way right along the structure. Paradoxically to make the structure as symmetrical as possible we had to have the two chains run in opposite directions; that is, the sequence of the atoms goes one way in one chain and the opposite way in the other.

Now we found that we could not arrange the bases any way we pleased; the four bases would fit into the structure only in certain pairs. In any pair there must always be one big one (purine) and one little one (pyrimidine). A pair of pyrimidines is too short to bridge the gap between the two chains, and a pair of purines is too big to fit into the space.

Adenine must always be pared with thymine and guanine with cytosine; it is impossible to fit the bases together in any other combination in our model. (This pairing is likely to be so fundamental for biology that I cannot help wondering whether some day an enthusiastic scientist will christen his newborn twins Adenine and Thymine!)

Now the exciting thing about a model of this type is that it immediately suggests how the DNA might produce an exact copy of itself. The model consists of two parts, each of which is the complement of the other. Thus, either chain may act as a sort of mold on which a complementary chain can be synthesized. The two chains

Vocabulary

helix らせん

wound wind（巻き付く）の過去分詞

pair off 2つずつ組む
paradoxically 逆説的に

please ～したいと思う

christen 名前をつける

complement 補完するもの

complementary chain 相補鎖

of a DNA, let us say, unwind and separate. Each begins to build a new complement onto itself. When the process is completed, there are two pairs of chains where we had only one. Moreover, because of the specific pairing of the bases the sequence of the pairs of bases will have been duplicated exactly; in other words, the mold has not only assembled the building blocks but has put them together in just the right order.

Vocabulary

unwind ほどける

specific 特異的な
duplicate 複製する

DNAが糖とリン酸基が交互につながった非常に長い鎖でできていることが現在では知られている。糖はどれもデオキシリボースという糖だ。このリン酸基と糖の鎖は完璧に規則的だが，分子全体はそうではない。それぞれの糖に「塩基」が付いているためだ。通常，4種の塩基が見られる。うち2つはアデニンとグアニンというプリン塩基であり，他の2つはチミンとシトシンというピリミジン塩基だ。これまでにわかったように，鎖に沿ってこれらの塩基が並んでいる順番は不規則であり，おそらくDNAの部分ごとに異なっている。DNAが鎖であることは化学式からわかっているものの，それだけでは分子の形はわからない。この鎖には多数の単結合が含まれ，この単結合を軸に回転が可能だから，鎖はどんな形にも巻き上がりうるだろう。

ケンブリッジ大学キャベンディッシュ研究所で研究しているワトソン（J. D. Watson）と私は，ロンドン大学キングスカレッジのウィルキンス（M. H. F. Wilkins）とフランクリン（Rosalind Franklin）らが撮影したX線パターンをもとにDNAのスケールモデルを作り，DNAの構造に近いものを手に入れたと確信した。何とか構造を推定するため，私たちはいくつかの仮定を置く必要があった。最も重要だったのは，結晶構造パターンの反復が鎖の化学組成の反復とは一致せず，もっと長い間隔で繰り返しているという事実に説明をつけることだった。考えうる可能性は，鎖を構成する輪はみな同じだが，X線で見えているのは10個の輪ごとに1個だけ，例えば同じ向きの輪で，その間にある向きの異なる輪は見えていないというものだ。どんな鎖ならこのパターンが生じるだろうか？　答えは簡単。この鎖は螺旋を巻いているのだ。結晶構造の反復距離は螺旋の一巻きから次の一巻きまでの間隔に対応しているのだろう。

この特徴的なモデルは，同じ軸に巻き付いた一対のDNA鎖を含んでいる。2本の鎖は塩基によって連結している。一方の鎖の塩基は他方の鎖の同じ階に位置する塩基と非常に弱い結合によって結びついており，すべての塩基がそのように構造全体にわたってペアをなしている。この構造をできるだけ対称的にするには，逆説的なことに，2本の鎖を逆向きに走らせる必要があった。つまり，一方の鎖の原子の配列が，他方の鎖とは逆に並んでいる。

ここで私たちは，塩基を好き勝手に並べるわけにはいかないことに気づいた。4種類の塩基は特定の相手と対になったときにのみ，この構造に落ち着くのだ。どの対も，片方の塩基は大きくて（プリン塩基），他方は小さなもの（ピリミジン塩基）でなくてはならない。ピリミジン塩基どうしのペアでは短すぎて2本の鎖の間を橋渡しできず，プリン塩基の対だと大きすぎて隙間に収まらない。

アデニンは常にチミンと，グアニンはシトシンと対になる必要がある。その他の組み合わせでは，私たちのモデルに決して合致しえない（このペアリングは生物学にとって非常に基本的だろうから，いつの日か熱心な科学者が自分のところに生まれた双子を「アデニンちゃん」「チミンちゃん」と名づけるのでは，と思わざるをえないほどだ）。

この種のモデルの素晴らしいところは，DNAが自分自身の正確なコピーをどのように作り出すのかをすぐに示唆してくれることだ。この分子モデルは2つの部分からなり，それぞれが他方の補完物になっている。だから，片方の鎖がいずれも一種の鋳型として働いて，その上に相補鎖を合成することができるだろう。1つのDNAを構成している2本の鎖がほどけ，分離したとしよう。それぞれの鎖が自分自身の上に新たな相補鎖を作り始める。この過程が完了すると，1対しかなかったものが2つになる。さらに，特定の塩基どうしが対になるため，塩基対の配列は正確に複製されるだろう。言い換えれば，この鋳型は材料を組み上げるだけでなく，それらを正しい順序にまとめ上げるのだ。

The Courtship of Animals

動物の求愛行動（1954 年掲載）

N. ティンバーゲン（1973 年受賞）

ニコラス・ティンバーゲン（Nikolaas Tinbergen, 1907 ～ 1988 年）はオランダの動物行動学者・鳥類学者。1973 年,「個体的および社会的行動様式の組織化と誘発に関する発見」によって，オーストリアの動物行動学者であるコンラート・ローレンツ（36 ページ参照）およびカール・フォン・フリッシュ（Karl von Frisch，1886 ～ 1982 年）とともにノーベル生理学・医学賞を受賞した。

ティンバーゲンは動物の行動が単に外部刺激に機械的に反応しているものではなく，動物自身の情動など内面に起因していると考えて，主に鳥類の行動を詳しく観察して考察を加えた。以下の記事はそのエッセンスを伝えている。

動物の行動はそれを専門としない一般市民にも興味深いものだが，ティンバーゲンやローレンツ，フォン・フリッシュはそれをきちんと科学的に整理してとらえる現代的な動物行動学を切り拓いた先駆者だ。

初出：SCIENTIFIC AMERICAN November 1954
抜粋掲載：SCIENTIFIC AMERICAN June 2011, 日経サイエンス 2011 年 11 月号

In contrast to such clearly motivated behavior as feeding or flight from predators, the courtship postures of animals are altogether puzzling, because it is difficult to see at first glance not only what circumstances cause them to occur but even what functions they serve. We may suppose that the male's display and activities stimulate the female to sexual cooperation, but even this elementary assumption has to be proved. And then we have to ask: Why does the female have to be stimulated in so elaborate a fashion, and what factors enter into the male's performance? Our work suggests that courtship serves not only to release sexual behavior in the partner but also to suppress contrary tendencies, that is, the tendencies to aggression or escape.

Let me give a brief sketch of what happens when gulls of the black-headed species form pairs at the beginning of the breeding season. An unmated male settles on a mating territory. He reacts to any other gull that happens to come near by uttering a "long call" and adopting an oblique posture. This will scare away a male, but it attracts females, and sooner or later one alights near him. Once she has alighted, both he and she suddenly adopt the "forward posture." Sometimes they may perform a movement known as "choking." Finally, after one or a few seconds, the birds almost simultaneously adopt the "upright posture" and jerk their heads away from each other. Now most of these movements also take place in purely hostile clashes between neighboring males. They may utter the long call, adopt the forward posture and go through the choking and the upright posture.

The final gestures in the courtship sequence—the partners' turning of their heads away from each other, or "head-flagging"—is different from the others: it

Vocabulary

courtship posture 求愛行動姿勢

display 求愛ディスプレー
▶ Technical Terms

elaborate 手の込んだ

aggression 攻撃
escape 逃避

breeding season 繁殖期
territory 縄張り

utter 声を発する
adopt 採用する
oblique 斜めの

alight 舞い降りる

jerk 急に動かす

hostile 敵対的な

Technical Terms

求愛ディスプレー（**display**） 動物が相手に自分の存在を誇示するために示す特定の姿勢や行為をディスプレーという。求愛のほか，威嚇などのディスプレーがあり，主に鳥類で研究が進んだ。

is not a threat posture. Sometimes during a fight between two birds we see the same head-flagging by a bird which is obviously losing the battle but for some reason cannot get away, either because it is cornered or because some other tendency makes it want to stay. This head-flagging has a peculiar effect on the attacker: as soon as the attacked bird turns its head away the attacker stops its assault or at least tones it down considerably. Head-flagging stops the attack because it is an "appeasement movement"—as if the victim were "turning the other cheek." We are therefore led to conclude that in their courtship these gulls begin by threatening each other and end by appeasing each other with a soothing gesture.

The black-headed gull is not an isolated case. We have learned that our courtship theory applies to many other birds (including various finches, tits, cormorants, gannets, ducks) and to animals of quite different groups, such as fish.

It is still an open question whether this gradual change in the motivational situation is mediated by endocrine changes, such as the growth of gonads. Future research will have to settle this. Our theory, as very briefly outlined here, is but a first step in the unraveling of the complicated causal relationships underlying the puzzling but fascinating phenomena of courtship.

Vocabulary

corner 隅に追い詰める

assault 攻撃

appeasement 慰撫, 宥和

soothe なだめる, 和らげる

tit シジュウカラ
cormorant ウ
gannet シロカツオドリ

open question 未解決の問題
endocrine 内分泌
gonad 生殖腺

causal relationship 因果関係

餌を食べたり捕食者から逃げたりといった明らかな動機のある行動とは対照的に，動物の求愛行動姿勢はまるでわけがわからない。一見したところでは，どういう状況が求愛姿勢を引き起こすのかだけでなく，求愛姿勢が何の機能を果たしているのかも判然としないからだ。雄の求愛ディスプレーと行動が雌の性的協力を刺激するのだと考えられるだろうが，この基本的な推定にも証明が必要だ。そこで，こう問わざるを得ない。雌はなぜ，これほど手の込んだやり方で刺激されなくてはならないのか，そして雄のパフォーマンスにはどういう要素が含まれているのか？　私たちの研究結果は，求愛行動がパートナーから性的行動を引き出すだけでなく，逆の傾向を抑制する，つまり攻撃や逃避の傾向を抑制する働きもあることを示している。

例として，ユリカモメの仲間が繁殖期の初めにペアを作る様子を手短に紹介しよう。１羽の独身の雄が繁殖縄張りにいる。彼は近くにやってきたカモメすべてに「長鳴き」で反応し，「傾斜姿勢」を取る。これは雄を脅して追い払う効果があるが，雌は惹きつけられ，遅かれ早かれ１羽の雌が彼の近くに舞い降りる。彼女が舞い降りるや，両者は突然「前傾姿勢」を取る。ときには「チョーキング」として知られる動きをすることもある。そして数秒もしないうちに，２羽はほとんど同時に「直立姿勢」を取り，互いに顔をそむける。さて，これらの動きの大部分は，隣どうしの雄の間に生じる敵対的衝突の際にも見られる。この場合も長鳴き，前傾姿勢，チョーキングを経て直立姿勢に至る。

パートナーが互いにそっぽを向く「顔そむけ（ヘッド・フラッギング）」という求愛行動の最後の動作は，ほかとは違って脅しの姿勢ではない。２羽の鳥が喧嘩しているとき，これと同じ顔そむけがときどき観察される。喧嘩に負けそうになったほうの鳥が，隅に追い詰められているためか，あるいは何か別の性向があってその場にとどまろうとするために逃げ出せない場合，顔そむけをするのだ。この動作は攻撃者に特異な効果を及ぼす。攻撃されている鳥が顔をそむけると，攻撃者は直ちに攻撃をやめるか，少なくともかなり弱めるのだ。顔そむけが攻撃を停止させるのは，これが「なだめの姿勢」であるからだ。ひどい仕打ちをされたのに，もう一方の頬を差し出して“仕返しをしない”といっているようなものである。したがって，これらのカモメは求愛行動において，互いを脅すことから始め，最後には慰撫の姿勢を取って互いをなだめているのだという結論に導かれる。

ユリカモメは特殊なケースではない。私たちのこの求愛行動仮説は他の数多くの鳥（フィンチやシジュウカラ，ウ，シロカツオドリ，カモなど）のほか，鳥とはまったく異なる魚などの動物にも当てはまることがわかっている

この動機づけ状況の漸進的な変化が生殖腺の成長など内分泌系の変化に影響を受けているのかどうかは未解決の問題だ。今後の研究によって明らかになるだろう。ここでは私たちの仮説をごく手短に紹介したが，これは求愛行動という不思議だが興味の尽きない現象の背後にある複雑な因果関係を解き明かす第一歩にすぎない。

Skin Transplants

皮膚移植（1957年掲載）

P. B. メダワー（1960年受賞）

ピーター・メダワー（Peter B. Medawar, 1915 ～ 1987年）はブラジル生まれの英国の生物学者。移植組織に対する免疫系の反応に関する研究で知られ，1960年に「後天的免疫寛容の発見」によってオーストラリアのウイルス学者フランク・バーネットとともにノーベル生理学・医学賞を受賞した（バーネットについては10ページ参照）。

メダワーは第二次世界大戦中に皮膚移植の研究を始め，組織移植とそれに対する免疫系の拒絶反応に関する理解を大きく進展させた。ノーベル賞を受賞する3年前に執筆された以下の記事は，その要点を解説している。

現在の医療では皮膚だけでなく内臓など複雑な臓器の移植も可能になっているが，拒絶反応が生じる仕組みは同じだ。メダワーらの研究は免疫抑制剤など拒絶反応を抑える手立ての開発につながった。

初出：SCIENTIFIC AMERICAN April 1957
抜粋掲載：SCIENTIFIC AMERICAN July 2014, 日経サイエンス 2014年11月号

Plainly the reaction against a graft is an immunological one; i.e., a reaction of the same general kind as that provoked in the body by foreign proteins, foreign red blood cells, or bacteria. This is easily demonstrated by experiments. After a mouse has received and rejected a transplant from another mouse, it will destroy a second graft from the same donor more than twice as rapidly, and in a way which shows that it has been immunologically forearmed. This heightened sensitivity is conferred upon a mouse even when it merely receives an injection of lymph node cells from a mouse that has rejected a graft.

In most immunological reactions the body employs antibodies as the destroying agent—e.g., in attacking foreign proteins, germs and so on. Antibodies are formed in response to a homograft (a transplant between different animals of the same species), but there are reasons to doubt that these are normally the instruments of the reaction against such a graft. Paradoxically enough, a high concentration of circulating antibodies seems if anything to weaken the reaction: it allows the graft to enjoy a certain extra lease of life.

The actual agents of attack on the graft seem to be not antibodies but cells produced by the lymph glands. Some skillfully designed experiments by G. H. Algire, J. M. Weaver and R. T. Prehn at the National Cancer Institute certainly do point in that direction.

In one experiment they enclosed a homograft in a porous capsule before planting it in a mouse which had been sensitized by an earlier homograft from the same donor. When the pores of the capsule were large enough to let cells through, the mouse destroyed the graft. But when

Vocabulary

plainly 明らかに
graft 移植組織
immunological 免疫学的な
provoke 引き起こす
transplant 移植したもの, 移植臓器, 移植組織
forearmed あらかじめ武装した
confer 授与する, 性質を与える
lymph node リンパ節
▶ Technical Terms

antibody 抗体
germ 病原菌, 微生物
homograft 同種移植片

concentration 濃度
if anything どちらかというと, むしろ
lease 人生の与えられた期間, 寿命

lymph gland リンパ腺
▶ Technical Terms

porous 多数の孔の開いた, 多孔質の

Technical Terms

リンパ節(**lymph node**)　全身からリンパ液を回収するリンパ管の途中にあって, 異物をチェックしている関所のような組織。
リンパ腺(**lymph gland**)　リンパ球を作り出している器官。胸腺など。

the experimenters used membranes with pores so fine that they kept out cells and let through only fluid, the graft survived.

The hypothesis that the action against a graft is carried out by cells explains why grafts in the cornea are mercifully exempted from attack. The cornea has no blood vessels; consequently blood-borne cells cannot reach the graft.

In the brain, on the other hand, the converse of this situation obtains: the brain lacks a lymphatic drainage system, so that any antigens released by a graft there may not be able to travel to centers where they can stir up an immunological response. This probably explains why homografts can often be transplanted successfully into the brain.

Vocabulary

fluid 体液

cornea 角膜
mercifully 幸いにも
be exempted from ～を免れる
consequently その結果
-borne ～によって運ばれる

obtain 存在する，通用する
drainage 排水，排水管
antigen 抗原
stir up かき立てる，強める

移植組織に対する反応は明らかに免疫学的な反応だ。外来タンパク質や外来の赤血球，細菌によって身体に引き起こされるのと総じて同じ種類の反応である。これは実験で容易に実証できる。他のマウスの組織を移植されたマウスがこれを拒絶した後に，同じドナーからの組織移植を繰り返すと，前回の2倍以上の速さで移植組織を破壊し，免疫的にあらかじめ武装していたことがわかる。この感受性の強化は，ドナーマウスのリンパ節細胞を注射しただけでも生じる。

多くの免疫反応において，身体は抗体を破壊要員として外来のタンパク質や病原菌などを攻撃する。抗体は同種移植片（同種の動物から移植された移植）に対しても形成されるが，これらが移植組織に対する拒絶反応の原因ではないと考えられる理由がある。逆説的なことに，高濃度の抗体が循環していると拒絶反応はむしろ弱まるようで，移植組織が生き長らえる期間がいくらか延びるのだ。

移植組織を実際に攻撃しているのは抗体ではなく，リンパ腺で生まれた細胞らしい。米国立がん研究所のアルジア（G. H. Algire）とウィーバー（J. M. Weaver），およびプレーン（R. T. Prehn）による巧みに設計されたいくつかの実験は，まさにこの線を示している。

彼らはマウスに別のマウスの組織を移植して感度を高めた後，同じドナーマウスからの同種移植片を多孔質のカプセルに入れて再び移植した。カプセルの孔が大きく，細胞が通過できる場合には，マウスは移植組織を破壊した。しかし，孔のサイズが小さな膜を使って細胞の通過を妨げ，体液の出入りだけを許した場合，移植組織は生き延びた。

移植片に対する反応が細胞によって行われているという仮説は，角膜に移植された組織片が幸いにも攻撃を免れる理由を説明する。角膜には血管がなく，したがって血液で運ばれる細胞が移植片に到達できないのだ。

一方，脳はこれと逆の状況になっている。脳にはリンパ液を外へ排出する流路がないので，組織移植によって脳で生まれた抗体は脳内にとどまり，免疫応答を強める免疫センターまで移動することができない。脳への同種移植片が成功することが多いのは，おそらくこのためだろう。

The Evolution of Behavior

行動の進化（1958 年掲載）

K. Z. ローレンツ（1973 年受賞）

コンラート・ツァハリアス・ローレンツ（Konrad Zacharias Lorenz，1903 ～ 1989 年）はオーストリアの動物行動学者。1973 年，「個体的および社会的行動様式の組織化と誘発に関する発見」でニコラス・ティンバーゲン，カール・フォン・フリッシュとともにノーベル生理学・医学賞を受賞した（27 ページ参照）。

ローレンツは「刷り込み」の研究者として知られ，ハイイロガンの雛に母親と間違われた体験がそのもとになったというエピソードはよく知られている。刷り込みだけでなく，動物の行動を厳密に観察することを基礎として，動物の行動とその進化を深く考察した。

ノーベル賞受賞の 15 年前に執筆された以下の記事は，個別の動物が示す行動と，その動物種に共通する生得的な行動パターンについて考察している。

初出：SCIENTIFIC AMERICAN December 1958
抜粋掲載：SCIENTIFIC AMERICAN June 2011，日経サイエンス 2011 年 11 月号

Following the example of zoologists, who have long exploited the comparative method, students of animal behavior have now begun to ask a penetrating question. We all know how greatly the behavior of animals can vary, especially under the influence of the learning process. But is it not possible that beneath all the variations of individual behavior there lies an inner structure of inherited behavior which characterizes all the members of a given species, genus or larger taxonomic group—just as the skeleton of a primordial ancestor characterizes the form and structure of all mammals today?

Yes, it is possible! Let me give an example which, while seemingly trivial, has a bearing on this question. Anyone who has watched a dog scratch its jaw or a bird preen its head feathers can attest to the fact that they do so in the same way. The dog props itself on the tripod formed by its haunches and two forelegs and reaches a hindleg forward in front of its shoulder. Now the odd fact is that most birds (as well as virtually all mammals and reptiles) scratch with precisely the same motion! A bird also scratches with a hindlimb (that is, its claw), and in doing so it lowers its wing and reaches its claw forward in front of its shoulder.

One might think that it would be simpler for the bird to move its claw directly to its head without moving its wing, which lies folded out of the way on its back. I do not see how to explain this clumsy action unless we admit that it is inborn. Before the bird can scratch, it must reconstruct the old spatial relationship of the limbs of the four-legged common ancestor which it shares with mammals.

Vocabulary

exploite 活用する
penetrating 鋭い，透徹した

species 種
▶ Technical Terms
genus 属
▶ Technical Terms
taxonomic group 分類群
mammal 哺乳類，哺乳動物

preen 羽づくろいをする
attest 証言する

haunches 動物の後躯

reptile 爬虫類

clumsy ぎこちない，不器用な
inborn 生まれつきの

Technical Terms

種（**species**）　生物を命名・分類するうえでの基本単位。厳密な定義をめぐっては専門家の間で議論があるが，相互に交配し合い，他の集団とは交配できない集団というのが伝統的な解釈だ。

属（**genus**）　生物の分類体系で種の上位に位置する階層。同類の種をまとめたグループと考えればよい。このほか一般に分類順序は，種ー属ー科（**family**）ー目（**order**）ー綱（**class**）ー門（**phylum, division**）ー界（**kingdom**）となる。

Comparative study of innate motor patterns represents an important part of the research program at the Max Planck Institute for Comparative Ethology. Our subjects are the various species of dabbling, or surface-feeding, ducks. By observing minute variations of behavior traits between species on the one hand and their hybrids on the other, we hope to arrive at a phylogenetics of behavior.

The first thing we wanted to know was how the courtship patterns of ducks become fixed. What happens when these ducks are crossbred? By deliberate breeding we have produced new combinations of motor patterns, often combining traits of both parents, sometimes suppressing the traits of one or the other parent and sometimes exhibiting traits not apparent in either. We have even reproduced some of the behavior-pattern combinations which occur in natural species other than the parents of the hybrid.

Thus, we have shown that the differences in innate motor patterns which distinguish species from one another can be duplicated by hybridization. This suggests that motor patterns are dependent on comparatively simple constellations of genetic factors.

Vocabulary

dabble 水底の餌を取るためにくちばしを突っ込む
surface-feeding 水面採食の
minute わずかな

phylogenetics 系統発生学
▶ Technical Terms

deliberate 意図的な

trait 形質
▶ Technical Terms

Technical Terms

系統発生学(**phylogenetics**) 生物種やそのグループが進化の歴史のなかでどのように生まれ分化してきたかの歴史を研究する学問。単に系統学ともいう。
形質(**trait**) 生物の分類の指標となる形態的特徴。また,遺伝的に表れる特徴全般のこと。

比較の手法を長らく用いてきた動物学者にならって，動物行動学の研究者たちは鋭い問いを発し始めた。動物の行動が，特に学習プロセスの影響下でいかにさまざまであるかは，よく知られている。しかし，それら個別のさまざまな行動の背景に，生物の種や属，さらに大きな分類群のメンバー全部に共通する，生得的行動の内部構造が存在するというのは，あり得ないことだろうか？　原始の祖先の骨格が現在の哺乳動物すべての体形と構造を特徴づけているのと同様に。

そう，その可能性はある！　この問題に関係する一見取るに足らない例を挙げよう。顎を掻いているイヌや，頭の羽を整えている鳥を観察したことのある人なら誰でも，彼らが同じやり方でそれを行っていると確言できるだろう。イヌは前肢２本と後躯で形成した３点で身体を支え，後ろ肢の１本を肩越しに前に伸ばして顎を掻く。ここで奇妙なのは，ほとんどの鳥が（実質的にすべての哺乳類と爬虫類も），これとまったく同じ動作で身体を掻くという事実だ！　鳥も後ろ肢（つまり爪のある足）を使って身体を掻き，その際には翼を下げて，爪のある後ろ肢を肩の前方へ持ってくる。

背中に畳んである翼をわざわざ動かすまでもなく，後ろ肢を頭へ直接持っていくほうが簡単なはずだと思えるだろう。これが生まれつきのものであると考えない限り，このぎこちない行動をどう説明すればよいか私にはわからない。鳥たちが頭を掻くには，哺乳類との四本足の共通祖先が持っていた四肢の空間的関係を前もって再現する必要があったのである。

生得的な運動パターンの比較研究はマックス・プランク動物行動学研究所での重要な研究分野をなしている。私たちの研究テーマはカモに見られるさまざまな水面採食行動だ。カモとその雑種の間に見られる行動習性のわずかな違いを観察することによって，私たちは動物行動の系統発生学に達したいと考えている。

最初に知りたかったのは，カモの求愛行動パターンがどのようにして現在のようなものになったのかということだ。そうしたカモが交雑するとどうなるか？　私たちは人工繁殖によって，新たな組み合わせの動作パターンを作り出した。たいていは両親の形質を組み合わせたものになるが，あるときにはどちらかの親の形質が抑えられ，またあるときは両親のどちらにもなかった形質が現れる。親とは別種のカモに見られる行動パターンの一部が，雑種に再現されることさえあった。

このように私たちは，生物種を区分している生得的動作パターンの違いを交雑によって再現できることを示した。これは，動作パターンが一連の比較的単純な遺伝因子によって決まることを示唆している。

Nerve Cells and Behavior

神経細胞と行動（1970年掲載）

E. R. カンデル（2000年受賞）

エリック・リチャード・カンデル（Eric Richard Kandel，1929年～）はオーストリア出身の米国の神経科学者。2000年，「神経系における情報伝達に関する発見」によって，スウェーデンの薬理学者アービド・カールソン（Arvid Carlsson，1923年～）および米国の神経科学者ポール・グリーンガード（Paul Greengard，1925年～）とともにノーベル生理学・医学賞を受賞した。

海生軟体動物アメフラシのニューロンに注目した実験で，CREBというタンパク質がないと長期記憶の形成が阻害されることを発見した。他の受賞者の研究も神経伝達物質やその生化学的プロセスに関する内容だ。

以下の記事は受賞の30年前に執筆されたもので，アメフラシの行動をニューロンという細胞レベルから理解しようとするアプローチについて述べている。その後，生化学的な詳細が実際に明らかになってきた。55ページ「ニューロンレベルでみた学習」も参照。

初出：SCIENTIFIC AMERICAN July 1970
抜粋掲載：SCIENTIFIC AMERICAN June 2011，日経サイエンス2011年11月号

Advances in the concepts and techniques for studying individual nerve cells and interconnected groups of cells have encouraged neural scientists to apply these methods to studying complete behavioral acts and modifications of behaviors produced by learning. This led to an interest in certain invertebrates, such as crayfish, leeches, various insects and snails, that have the great advantage that their nervous system is made up of relatively few nerve cells (perhaps 10,000 or 100,000 compared with the trillion or so in higher animals). In these animals one can begin to trace, at the level of individual cells, not only the sensory information coming into the nervous system and the motor actions coming out of it but also the total sequence of events that underlies a behavioral response.

The most consistent progress has come from studies of habituation and dishabituation in the spinal cord of the cat and the abdominal ganglion of *Aplysia* [a giant marine snail that grows to about a foot in size].

Habituation is a decrease in a behavioral response that occurs when an initially novel stimulus is presented repeatedly. Once a response is habituated, two processes can lead to its restoration. One is spontaneous recovery, which occurs as a result of withholding the stimulus to which the animal has habituated. The other is dishabituation, which occurs as a result of changing the stimulus pattern, for example, by presenting a stronger stimulus to another pathway.

An *Aplysia* shows a defensive withdrawal response [to gentle stimulation]. The snail's gill, an external respiratory organ, is partially covered by the mantle shelf, which

Vocabulary

invertebrate 無脊椎動物
crayfish ザリガニ
leech ヒル
nervous system 神経系

habituation 馴化
▶ Technical Terms
dishabituation 脱馴化
spinal cord 脊髄
abdominal ganglion 腹部神経節
Aplysia アメフラシ

retoration 回復, 復活
withhold 差し控える, 与えずにおく

gill エラ
respiratory 呼吸の

Technical Terms

馴化(**habituation**)　生物が示す適応や順応(**adaptation**)のこと。本文の文脈における意味は順応に近い。「順化」とも書く。

contains the thin residual shell. When either the mantle shelf or anal siphon, a fleshy continuation of the mantle shelf, is gently touched, the siphon contracts and the gill withdraws into the cavity under the mantle shelf.

We can now propose a simplified circuit diagram to illustrate the locus and mechanism of the various plastic changes that accompany habituation and dishabituation of the gill-withdrawal reflex. Repetitive stimulation of sensory receptors leads to habituation by producing a plastic change at the synapse between the sensory neuron and the motor neuron. Stimulation of the head leads to dishabituation by producing heterosynaptic facilitation at the same synapse.

It would seem that cellular approaches directed toward working out the wiring diagram of behavioral responses can now be applied to more complex learning processes.

Vocabulary

anal siphon 吸管
fleshy 肉の，肉質の

plastic change 可塑的変化

reflex 反射
sensory receptor 感覚受容器
▶ Technical Terms
synapse シナプス
▶ Technical Terms
sensory neuron 感覚ニューロン
motor neuron 運動ニューロン
heterosynaptic facilitation 異シナプス促通

Technical Terms

感覚受容器（**sensory receptor**）　受容器（**receptor**）は動物が外界からの刺激情報の受け入れ口として備えている構造の総称で，感覚受容器は光や機械的な力など感覚に関する刺激に対応した受容器のこと。
シナプス（**synapse**）　ニューロン（神経細胞）どうしやニューロンと他の細胞を接合して信号伝達に関わっている構造。ニューロンの軸索の終末が次の細胞と狭い隙間を隔てて接しており，シナプス間隙と呼ばれるこの隙間に神経伝達物質が放出され，これが受け手のニューロンの受容体に結合することで信号が伝えられる。

個々の神経細胞や細胞グループ間の連絡を調べる技法が進歩したことで，行動や，学習によって生じる行動の変化を研究するのにそうした手法を適用する機運が神経科学者の間で高まってきた。この結果，ザリガニやヒル，さまざまな昆虫，貝類など，一群の無脊椎動物に対する関心が高まった。これらの無脊椎動物は神経細胞の数が比較的少ない（高等動物が1兆個程度なのに対し，おそらく1万～10万個ほど）という大きな利点がある。これらの動物では，神経系から入力される感覚情報やそこから生じる運動を個々の細胞レベルで追跡するだけでなく，行動応答の背後にある一連の出来事すべてを追えるのだ。

特に一貫して進んできたのは，ネコの脊髄とアメフラシ（30cmほどになる大きな海生軟体動物）の腹部神経節における馴化（じゅんか）と脱馴化の研究だ。

馴化とは，最初は新規だった刺激が繰り返し与えられたときに生じる行動反応の衰えだ。馴化が起こった場合，それが元に戻るには2つの道筋がある。1つは自発的回復で，その動物に当の馴化刺激を与えるのを控えておくと自然に起こる。もう1つが脱馴化であり，例えばある強い刺激を別の経路に与えるなど，刺激のパターンを変えた結果として起こる。

アメフラシは穏やかな刺激に対して「引き込み応答」という防御行動を示す。この動物の外部呼吸器官であるエラは一部が外被に覆われており，この外被には薄い殻がある。外被や，外被の延長である肉質組織の吸管にそっと触れると，吸管が縮み，エラは外被内部の空間に引っ込む。

このエラ引き込み反射の馴化と脱馴化に伴うさまざまな可塑的変化のメカニズムと，それが起きている場所について，いまや私たちはそれを示す単純な回路図を提案できる。感覚受容器を繰り返し刺激すると，感覚ニューロンと運動ニューロンの間のシナプスに可塑的変化が生じ，馴化につながる。頭部を刺激すると，その同じシナプスで異シナプス促通（興奮の伝達が起こりやすくなること）が生じ，脱馴化につながる。

行動応答の配線図を読み解くのに使われた細胞レベルのアプローチを，今度はもっと複雑な学習プロセスに適用することが可能だろう。

The Immune System

免疫系の機能（1973 年掲載）

N. K. イェルネ（1984 年受賞）

ニールス・カイ・イェルネ（Niels Kaj Jerne，1911 ～ 1994 年）はロンドン生まれのデンマークの免疫学者。1984 年に「免疫系の発達と制御における選択性に関する諸理論およびモノクローナル抗体の作成原理の発見」によって，ドイツの生物学者ジョルジュ・ケーラー（Georges J. F. Köhler，1946 ～ 1995 年）およびアルゼンチン出身の英国の生化学者セーサル・ミルスタイン（César Milstein, 1927 ～ 2002 年）とともにノーベル生理学・医学賞を受賞した。

ジフテリア抗毒素に関する研究など免疫関連で多くの業績がある。そうした個別の研究もさることながら，「人体に存在する抗体分子は互いに反応し合いながら 1 つのネットワークを形成している」とする「ネットワーク説」を提唱し，これがノーベル賞受賞に結びついた面が大きい（ちなみに授賞理由にある「モノクローナル抗体の作成原理の発見」はケーラーとミルスタインの業績）。

受賞の 10 年ほど前に執筆された以下の記事でも，膨大な数の抗体分子やリンパ球が相互作用するネットワークとしての免疫系が強調されている。

初出：SCIENTIFIC AMERICAN July 1973, サイエンス 1973 年 9 月号
抜粋掲載：SCIENTIFIC AMERICAN July 2014, 日経サイエンス 2014 年 11 月号

The immune system is comparable in the complexity of its functions to the nervous system. Both systems are diffuse organs that are dispersed through most of the tissues of the body. In man the immune system weighs about two pounds. It consists of about a trillion cells called lymphocytes and about 100 million trillion molecules called antibodies that are produced and secreted by the lymphocytes. The special capability of the immune system is pattern recognition, and its assignment is to patrol the body and guard its identity.

The cells and molecules of the immune system reach most tissues through the bloodstream, entering the tissues by penetrating the walls of the capillaries. After moving about, they make their way to a return vascular system of their own, the lymphatic system. The tree of lymphatic vessels collects lymphocytes and antibodies, along with other cells and molecules and the interstitial fluid that bathes all the body's tissues, and pours its contents back into the bloodstream by joining the subclavian veins behind the collarbone.

Lymphocytes are found in high concentrations in the lymph nodes, way stations along the lymphatic vessels, and at the sites where they are manufactured and processed: the bone marrow, the thymus and the spleen. The immune system is subject to continuous decay and renewal. During the few moments it took you to read this far, your body produced 10 million new lymphocytes and a million billion new antibody molecules. This might not be so astonishing if all these antibody molecules were identical. They are not. Millions of different molecules are required to cope with the task of pattern recognition, just as millions of different keys are required to fit millions of

Vocabulary

immune system 免疫系
nervous system 神経系

lymphocyte リンパ球
antibody 抗体
pattern recognition パターン認識
assignment 任務

capillary 毛細血管
vascular system 脈管系
lymphatic system リンパ系

interstitial fluid 間質液
▶ Technical Terms

subclavian veins 鎖骨下静脈
collarbone 鎖骨

lymph node リンパ節
lymphatic vessel リンパ管

bone marrow 骨髄
thymus 胸腺
▶ Technical Terms
spleen 脾臓

Technical Terms

間質液（**interstitial fluid**）　生体組織中で細胞を取り巻いている液体。血管やリンパ管を流れているものを除く体液。
胸腺（**thymus**）　胸腔にあるリンパ器官でＴ細胞など免疫細胞の生成に関与している。

different locks. The specific patterns that are recognized by antibody molecules are epitopes: patches on the surface of large molecules such as proteins, polysaccharides and nucleic acids. Molecules that display epitopes are called antigens. It is hardly possible to name a large molecule that is not an antigen.

The immune system and the nervous system are unique among the organs of the body in their ability to respond adequately to an enormous variety of signals. Both systems display dichotomies: their cells can both receive and transmit signals, and the signals can be either excitatory or inhibitory.

The nerve cells, or neurons, are in fixed positions in the brain, the spinal cord and the ganglia, and their long processes, the axons, connect them to form a network. The ability of the axon of one neuron to form synapses with the correct set of other neurons must require something akin to epitope recognition. Lymphocytes are 100 times more numerous than nerve cells and, unlike nerve cells, they move about freely. They too interact, however, either by direct encounters or through the antibody molecules they release. These elements can recognize as well as be recognized, and in so doing they too form a network. As in the case of the nervous system, the modulation of the network by foreign signals represents its adaptation to the outside world. Both systems thereby learn from experience and build up a memory, a memory that is sustained by reinforcement but cannot be transmitted to the next generation. These striking analogies in the expression of the two systems may result from similarities in the sets of genes that encode their structure and that control their development and function.

Vocabulary

epitope エピトープ
▶ Technical Terms
polysaccharide 多糖類
nucleic acid 核酸
antigen 抗原

dichotomy 二分法
excitatory 刺激性の
inhibitory 抑制性の

neuron ニューロン, 神経細胞
spinal cord 脊髄
ganglion 神経節。gangliaは複数形
process 突起
axon 軸索
synapse シナプス

modulation 調整, 変調

reinforcement 補強, 増強

Technical Terms

エピトープ(**epitope**)　抗体によって認識される抗原分子の部分構造のこと。抗体分子はここに結合する。「抗原決定基」ともいう。

免疫系は，その機能の複雑さにおいて，神経系に匹敵する。どちらの系も全身のほとんどの組織に分散して存在する臓器である。ヒトの免疫系の重さは約 900g あり，約 1 兆個のリンパ球と，リンパ球が産出し分泌する約 1 垓（10^{20}）個の抗体分子から成る。免疫系に特有な働きはパターン識別であり，その任務は全身をパトロールしてそれを守ることにある。

免疫系の細胞と分子は血流を通じてほとんどの組織に到達し，毛細血管壁を透過して組織へ入り込む。付近をうろついた後，独自の帰り道であるリンパ系に入る。リンパ管の枝はリンパ球と抗体を集めると同時に，他の細胞や分子，全身の組織をひたしている間質液も集め，鎖骨の後ろで鎖骨下静脈に合流して，その内容を血流の中へ戻す。

リンパ球はリンパ管の途中にあるリンパ節に高濃度に見られるほか，リンパ球を産生し処理する骨髄や胸腺，脾臓にも多く存在する。免疫系は絶え間なく壊れていく一方で再生されている。読者がここまで読み進めてきたわずかな間にも，あなたの身体は 1000 万個の新しいリンパ球と 1000 兆個の新しい抗体分子を生み出している。これらの抗体分子がすべて同じならそれほど驚くにはあたらないかもしれないが，実はそうではない。数百万個の違った錠前を開けるには数百万個の別々の鍵が必要なように，パターン識別の任務を遂行するには何百万もの異なる分子が必要だ。抗体分子によって識別される特異なパターンをエピトープといい，タンパク質や多糖類，核酸などの大きな分子の表面にある特定の一部分がこれである。このエピトープを持つ分子は抗原と呼ばれる。大きな分子で抗原にならないものはないといってよい。

免疫系と神経系は，非常に多様な信号に対してうまく反応する能力を持っている点で，身体の臓器の中では特異な存在だ。どちらの系も 2 つの対極的な性質を併せ持っている。その細胞は信号を受け取るとともに他の細胞に伝達もしており，その信号は刺激性のものと抑制性のものがある。

神経細胞（ニューロン）は脳や脊髄，神経節にあって固定した位置を保ち，その長い突起（軸索）で互いに連絡してネットワークを形成している。1 つのニューロンの軸索が他の適切な一連のニューロンとシナプスを作るには，エピトープ識別に類した何かが必要に違いない。リンパ球の数は神経細胞の 100 倍もあり，神経細胞とは違って自由に動き回る。しかしリンパ球は，直接出合うか，放出した抗体分子を通じてのいずれかによって，やはり相互作用している。これらの要素は相手から識別されるとともに相手を識別することができ，そうすることによって，やはりネットワークを形成している。神経系の場合と同様，外からの信号によるネットワークの調整が，外界への適応を示す。したがって，両系とも経験から学び，記憶を形成している。補強によって維持されるが，次の世代に伝達することはできない記憶だ。2 つの系の表現法がこのように非常によく似ているのは，その構造をコード化し発達と機能を調節している遺伝子群の類似から生じているのかもしれない。

Brain Mechanisms of Vision

視覚の脳内機構（1979年掲載）

D. H. ヒューベル／ T. N. ウィーセル（ともに1981年受賞）

デイヴィッド・ハンター・ヒューベル（David Hunter Hubel，1926～2013年）はカナダ出身の米国の神経生理学者，トルステン・ニルズ・ウィーセル（Torsten Nils Wiesel，1924年～）は米国で活動するスウェーデンの神経科学者で，ともに「視覚系における情報処理に関する発見」によって1981年のノーベル生理学・医学賞を受賞した。ちなみに同年の生理学・医学賞の残り半分は「大脳半球の機能分化に関する発見」によって米国の神経心理学者ロジャー・スペリー（Roger W. Sperry，1913～1994年）に贈られた。

ヒューベルとウィーセルは脳の視覚情報処理を実験によって調べた先駆者だ。1959年の実験で，ネコの視覚野に微小電極を挿入し，スクリーンに映した明暗のパターンを見せて反応を調べた。この結果，ある特定の方位（傾き）の線に対して反応するニューロンが別の角度の線には反応しないことを発見した。その後，特定の位置や方位に反応するニューロンが組み合わさって視覚野を形成し，個別の刺激が像として認識される仕組みが明らかになった。

ノーベル賞受賞の2年前に執筆された以下の記事はこの研究の要点を解説している。

初出：SCIENTIFIC AMERICAN September 1979, サイエンス 1979年11月号
抜粋掲載：SCIENTIFIC AMERICAN July 2014, 日経サイエンス 2014年11月号

The cerebral cortex, a highly folded plate of neural tissue about two millimeters thick, is an outermost crust wrapped over the top of, and to some extent tucked under, the cerebral hemispheres. In this article we hope to sketch the present state of knowledge of one subdivision of the cortex: the primary visual cortex, the most elementary of the cortical regions concerned with vision.

We can best begin by tracing the visual path in a primate from the retina to the cortex. The output from each eye is conveyed to the brain by about a million nerve fibers bundled together in the optic nerve. These fibers are the axons of the ganglion cells of the retina. A large fraction of the optic-nerve fibers pass uninterrupted to two nests of cells deep in the brain called the lateral geniculate nuclei, where they make synapses. The lateral geniculate cells in turn send their axons directly to the primary visual cortex.

To examine the workings of this visual pathway our strategy since the late 1950s has been (in principle) simple. Beginning, say, with the fibers of the optic nerve, we record with microelectrodes from a single nerve fiber and try to find out how we can most effectively influence the firing by stimulating the retina with light. For this one can use patterns of light of every conceivable size, shape and color, bright on a dark background or the reverse, and stationary or moving. Working in this way, one finds that both a retinal ganglion cell and a geniculate cell respond best to a roughly circular spot of light of a particular size in a particular part of the visual field.

The first of the two major transformations accomplished by the visual cortex is the

Vocabulary

cerebral cortex 大脳皮質
cerebral hemisphere 大脳半球

primary visual cortex 一次視覚野
▶ Technical Terms

primate 霊長類
retina 網膜
nerve fiber 神経線維
optic nerve 視神経
axon 軸索
ganglion cell 網膜神経節細胞
nest 集団
lateral geniculate nucleus 外側膝状体
synapse シナプス

microelectrode 微小電極

firing (ニューロンの)発火

transformation 変容, 変化

Technical Terms

一次視覚野(**primary visual cortex**) 視覚野は大脳皮質のうち視覚の処理に関わっている領域で, 後頭葉にある。一次視覚野(V1)は視覚処理の最初の段階を担っている部分で, 他のより高次な処理を行う脳領域に情報を送っている。

rearrangement of incoming information so that most of its cells respond not to spots of light but to specifically oriented line segments. There is a wide variety of cell types in the cortex, some simpler and some more complex in their response properties, and one soon gains an impression of a kind of hierarchy, with simpler cells feeding more complex ones. A typical cell responds only when light falls in a particular part of the visual world. The best response is obtained when a line that has just the right tilt is flashed in the region or, in some cells, is swept across the region. The most effective orientation varies from cell to cell and is usually defined sharply enough so that a change of 10 or 20 degrees clockwise or counterclockwise reduces the response markedly or abolishes it. (It is hard to convey the precision of this discrimination. If 10 to 20 degrees sounds like a wide range, one should remember that the angle between 12 o'clock and one o'clock is 30 degrees.)

There was a time, not so long ago, when one looked at the millions of neurons in the various layers of the cortex and wondered if anyone would ever have any idea of their function. For the visual cortex the answer seems now to be known in broad outline: Particular stimuli turn neurons on or off; groups of neurons do indeed perform particular transformations. It seems reasonable to think that if the secrets of a few regions such as this one can be unlocked, other regions will also in time give up their secrets.

Vocabulary

hierarchy 階層構造

tilt 傾き

markedly 著しく

abolish 廃する，完全に破壊する

give up (秘密を)明かす

大脳皮質は厚さ約2mmの神経組織の層で，大脳半球の最外層をなし，一部は折りたたまれた形で大脳半球の下にも回り込んでいる。この小論では，大脳皮質の一部であり，視覚に関係する皮質領野の中で最も基本的な一次視覚野について，現在得られている知識を概観したいと思う。

まず霊長類の視覚伝導路を網膜から大脳皮質までたどることから始めよう。それぞれの眼からの出力は，100万本ほどの神経線維の集まりである視神経によって脳に運ばれる。この線維は網膜神経節細胞の軸索に当たる。視神経線維の大部分は，そのまま脳の深部にある「外側膝状体」と呼ばれる2つの細胞集団に達し，そこでシナプスを形成する。そして外側膝状体の細胞の軸索は一次視覚野に直接達している。

私たちは1950年代後半から視覚伝導路の働きを研究してきたが，そこで用いた方法は（原則としては）単純なものだ。まず，微小電極を用いて単一の視神経線維の活動を記録しながら，網膜にどんな光刺激を加えたらその線維の発火が最も強く影響を受けるかを調べる。様々な大きさや形，色のパターンを，暗い背景に明るい図として，または逆に明るい背景の上に暗い図として示し，それらを静止あるいは動かした状態で提示する。こうして調べた結果，網膜神経節細胞も外側膝状体の細胞も，視野の特定部分に特定の大きさをしたほぼ円形の明るいスポットが呈示された時に最もよく反応することがわかった。

皮質視覚野では2つの大きな変化が起こるが，その1つは入力情報が再編されて，視覚野の細胞のほとんどが光のスポットにではなく，特定の方位の線に反応するようになることだ。視覚野には反応特性が単純なものから復雑なものまで様々なタイプの細胞があり，単純な細胞からの出力が複雑な細胞への入力となっているような，一種の階層構造がうかがわれる。ある典型的な細胞は，視野内の特定の位置に光が当たったときにだけ反応する。またある細胞は，特定の位置に特定の傾きを持つ線が示された時に，また一部の細胞ではそうした線が領域を横切って移動する時に反応する。最も有効な線の方位は細胞によって異なるが，その幅は狭く，時計回りか反時計回りに10～20°回転しただけで，反応は著しく減少または消失する（この判別力の正確さを理解してもらうのはなかなか難しい。10～20°と聞いて大雑把だと思う読者は，時計の12時と1時の間の角度が 30°であることを思い出してほしい)。

大脳皮質の各層に散らばる何百万もの細胞を目のあたりにして，それらの機能について何らかの考えを思いつくことが果たして可能だろうかと思われていたのは，そう昔のことではない。それが今日では，視覚皮質に関しては，その答えの概要が見えてきたようだ。特定の刺激がニューロンをオン・オフし，一群のニューロンが実際に特定の情報再編を実行しているのである。視覚野など少数の領野で秘密が解き明かされれば，他の領域の秘密もいずれは秘密でなくなるだろう。

The Biological Basis of Learning and Individuality

ニューロンレベルでみた学習（1992年掲載）

E. R. カンデル（2000年受賞）／R. D. ホーキンス

カンデルは2000年，「神経系における情報伝達に関する発見」によってノーベル生理学・医学賞を共同受賞した（41ページ参照）。

彼は「神経細胞と行動」（1970年掲載）で述べたアメフラシの実験からさらに踏み込み，記憶が脳のなかでどのように形成されるのかをニューロンのレベルに立って理解しようと研究を続けた。ノーベル賞受賞の8年前に執筆されたこの記事は，当時の最新の知見をまとめている。1970年に比べ，より詳細な事実が明らかになっている。共著者のホーキンス（Robert D. Hawkins）はコロンビア大学神経科学科の教授（執筆当時は准教授）。

もっとも，学習や記憶など，脳の高次機能のメカニズムはいまなお未解明であり，21世紀の科学が挑む大きな謎といえる。

初出：SCIENTIFIC AMERICAN September 1992, 日経サイエンス 1992年11月号
抜粋掲載：SCIENTIFIC AMERICAN July 2014, 日経サイエンス 2014年11月号

Elementary aspects of the neuronal mechanisms important for several different types of learning can now be studied on the cellular and even on the molecular level. Researchers agree that [some] forms of learning and memory require a conscious record. These types of learning are commonly called declarative or explicit. Those forms of learning that do not utilize conscious participation are referred to as nondeclarative or implicit.

Explicit learning is fast and may take place after only one training trial. It often involves association of simultaneous stimuli and permits storage of information about a single event that happens in a particular time and place; it therefore affords a sense of familiarity about previous events. In contrast, implicit learning is slow and accumulates through repetition over many trials. It often involves association of sequential stimuli and permits storage of information about predictive relations between events. Implicit learning is expressed primarily by improved performance on certain tasks without the subject being able to describe just what has been learned, and it involves memory systems that do not draw on the contents of the general knowledge of the individual.

The existence of two distinct forms of learning has caused the reductionists among neurobiologists to ask whether there is a representation on the cellular level for each of these two types of learning process. Canadian psychologist Donald O. Hebb boldly suggested that associative learning could be produced by a simple cellular mechanism. He proposed that associations could be formed by coincident neural activity: "When an axon of cell A ... excite[s] cell B and repeatedly or persistently takes part in firing it, some growth process or metabolic

Vocabulary

declarative 陳述的
▶ **Technical Terms**
explicit 顕在的
nondeclarative 非陳述的
implicit 潜在的
afford 提供する，与える
repetition 繰り返し，反復
predictive 予言的な
subject 本人
reductionist 還元主義者
boldly 大胆に
association 連合学習
axon 軸索

Technical Terms

陳述記憶（**declarative memory**）　教科書で学んだ知識や体験した出来事など，内容を言葉で表現できる記憶のこと。「宣言的記憶」ともいう。

change takes place in one or both cells such that A's efficacy, as one of the cells firing B, is increased." According to Hebb's learning rule, coincident activity in the presynaptic and postsynaptic neurons is critical for strengthening the connection between them (a so-called pre-post associative mechanism).

Ladislav Tauc and one of us (Kandel) proposed a second associative learning rule in 1963 while working at the Institute Marey in Paris on the nervous system of the marine snail *Aplysia*. They found that the synaptic connection between two neurons could be strengthened without activity of the postsynaptic cell when a third neuron acts on the presynaptic neuron. The third neuron, called a modulatory neuron, enhances transmitter release from the terminals of the presynaptic neuron. They suggested that this mechanism could take on associative properties if the electrical impulses known as action potentials in the presynaptic cell were coincident with action potentials in the modulatory neuron (a pre-modulatory associative mechanism).

Subsequently, we and our colleagues found experimental confirmation. We observed the pre-modulatory associative mechanism in *Aplysia*, where it contributes to classical conditioning, an implicit form of learning. Then, in 1986, Holger J. A. Wigström and Bengt E. W. Gustafsson, working at the University of Göteborg, found that the pre-post associative mechanism occurs in the hippocampus, where it is utilized in types of synaptic change that are important for spatial learning, an explicit form of learning.

Vocabulary

Hebb's learning rule ヘッブの学習則
presynaptic neuron シナプス前ニューロン
▶ Technical Terms
postsynaptic neuron シナプス後ニューロン

Aplysia アメフラシ

modulatory neuron 調節ニューロン
transmitter 神経伝達物質
▶ Technical Terms

action potential 活動電位

conditioning 条件づけ

hippocampus 海馬

spatial learning 空間学習

Technical Terms

シナプス前ニューロン(**presynaptic neuron**)　2つのニューロンが接続するシナプスにおいて,信号の送り手側となるニューロンのこと。これに対し,信号の受け手となるのがシナプス後ニューロン。

神経伝達物質(**neurotransmitter**)　シナプス前ニューロンからシナプス間隙に放出される化学物質で,これをシナプス後ニューロンの受容体がキャッチすることで信号が伝わる。神経伝達物質には様々なものがある。

異なるタイプの学習に重要な神経機構の基本的な側面を，細胞レベル・分子レベルで研究することが可能になっている。ある形態の学習と記憶は意識的な記録を必要とし，そうしたタイプの学習は「陳述的」あるいは「顕在的」と呼ばれている。一方，本人の意識的な参与を用いない形態の学習は「非陳述的」あるいは「潜在的」と呼ばれる。

顕在的学習は迅速で，たった1回の訓練で達成されることもある。また，同時に生じた刺激の関連づけを必要とすることが多く，特定の時と場所で起きた単一の出来事に関する情報が記憶される。このため以前の出来事について「知っている」という感覚が生じる。これに対し潜在的学習はゆっくりと進み，何回もの試行の繰り返しを通じて蓄積される。引き続いて起こった一連の刺激の関連づけを必要とする場合が多く，事象間に予想される関係についての情報が記憶される。ある課題を遂行するのが上手になったものの，何を学習したから上手になったのかを本人は説明できないというのが潜在的学習の典型であり，当人の一般的知識内容によらない記憶系が関与している。

このように学習に明らかに異なる2つの型があることから，神経生物学者のうち還元主義的な人たちは「それぞれの学習過程に対応する細胞レベルの現象が存在するのではないか」と考えた。カナダの心理学者ヘッブ（Donald O. Hebb）は大胆にも「連合学習は単純な細胞レベルのメカニズムによって生じうる」と考え，時間的に一致した神経活動によって関連づけが形成されると提唱した。具体的には「ニューロンAの軸索がニューロンBを刺激し，繰り返しあるいは持続的にその発火に関与すると，何らかの成長あるいは代謝の変化が一方または両方のニューロンに生じ，AがBを活性化する効率が増強する」というものだ。この「ヘッブの学習則」によれば，シナプス前ニューロンとシナプス後ニューロンが同時に活性化することが，そのシナプス結合の増強に必須となる（いわゆる「シナプス前-後連合メカニズム」）。

トーク（Ladislav Tauc）と著者の1人カンデルは1963年，第2の連合学習則を提案した。2人は当時，パリのマレー研究所で海生腹足類のアメフラシの神経系を研究するなかで，シナプス後ニューロンの活動を伴わなくても，第3の細胞がシナプス前細胞に作用するとシナプス結合が強まる場合があることを発見した。この第3の細胞は「調節ニューロン」と呼ばれ，シナプス前ニューロンからの神経伝達物質の放出を促進する。シナプス前ニューロンの電気インパルス（活動電位）が調節ニューロンの活動電位と同期している場合，このメカニズムは関連づけの性質を持ちうると2人は提案した（シナプス前 - 調節性連合メカニズム）。

その後の実験でこの仮説は確認された。私たちはアメフラシにおけるシナプス前 - 調節性連合機構が潜在的学習の一種である古典的条件づけに関与していることを観察した。1986年，イエーテボリ大学のウィグストレム（Holger W. A. Wigström）とグスタフソン（Bengt E. W. Gustafsson）はシナプス前 - 後連合メカニズムが海馬において，空間学習に重要なシナプス変化を生じるのに関与していることを発見した。空間学習は顕在的学習の一種だ。

The Prion Diseases

プリオン病はどこまで解明されたか（1995年掲載）

S. B. プルシナー（1997年受賞）

スタンリー・ベン・プルシナー（Stanley Ben Prusiner，1942年〜）は米国の生化学者。1997年，「プリオン－感染症の新たな生物学的原理－の発見」でノーベル生理学・医学賞を単独で受賞した。

プリオンは病原性物質として振る舞うタンパク質で，ウシ海綿状脳症（BSE, いわゆる狂牛病）の原因として一般市民にもよく知られるようになった。同種の疾患である羊のスクレイピーや人間のクロイツフェルト・ヤコブ病（CJD）の原因でもある。

それまで感染症の病原体といえば細菌やウイルスなど遺伝物質を備えた“生き物”であるというのが常識だったため，単なるタンパク質が病原体であるというプルシナーの見解は当初は懐疑的に受け止められた。ノーベル賞受賞2年前に執筆された以下の記事も，そうした初期の反発を回顧している。

また，抜粋記事の最後にプリオンが「アルツハイマー病など一般的な神経変性疾患に関係しているかどうかはまだ不明だが，その可能性を無視すべきではない」と述べられているが，実際これまでに「その可能性」を示唆する研究結果が出てホットなテーマとなっている。

初出：SCIENTIFIC AMERICAN January 1995，日経サイエンス1995年3月号
抜粋掲載：SCIENTIFIC AMERICAN June 2011，日経サイエンス2011年11月号

Fifteen years ago I evoked a good deal of skepticism when I proposed that the infectious agents causing certain degenerative disorders of the central nervous system in animals and, more rarely, in humans might consist of protein and nothing else. At the time, the notion was heretical. Dogma held that the conveyers of transmissible diseases required genetic material, composed of nucleic acid (DNA or RNA), to establish an infection in a host. Even viruses, among the simplest microbes, rely on such material to direct synthesis of the proteins needed for survival and replication. Later, many scientists were similarly dubious when my colleagues and I suggested that these "proteinaceous infectious particles"—or "prions," as I called the disease-causing agents—could underlie inherited, as well as communicable, diseases. Such dual behavior was then unknown to medical science. And we met resistance again when we concluded that prions (pronounced "PREE-eons") multiply in an incredible way; they convert normal protein molecules into dangerous ones simply by inducing the benign molecules to change their shape. Today, however, a wealth of experimental and clinical data has made a convincing case that we are correct on all three counts.

The known prion diseases, all fatal, are sometimes referred to as spongiform encephalopathies. They are so named because they frequently cause the brain to become riddled with holes. These ills, which can brew for years (or even for decades in humans), are widespread in animals. The most common form is scrapie, found in sheep and goats. Mad cow disease is the most worrisome. [The human prion diseases include among them Creutzfeldt-Jakob disease, a cause of dementia.]

Vocabulary

central nervous system 中枢神経系
protein タンパク質
heretical 異端の
nucleic acid 核酸
replication 複製，増殖
prion プリオン
▶ Technical Terms
incredible 信じがたい
spngeform encephalopathy 海綿状脳症
brew 醸造される，（いやなことが）準備される
scrapie スクレイピー
Creutzfeldt-Jakob disease クロイツフェルト・ヤコブ病
dementia 認知症

Technical Terms

プリオン（**prion**）　タンパク質でできた感染性因子。病原性のプリオンは周囲の正常なタンパク質の立体構造を変えてやはり病原性にしていく。

In addition to showing that a protein can multiply and cause disease without help from nucleic acids, we have gained insight into how scrapie PrP ["prion protein"] propagates in cells. Many details remain to be worked out, but one aspect appears quite clear: the main difference between normal PrP and scrapie PrP is conformational. Evidently, the scrapie protein propagates itself by contacting normal PrP molecules and somehow causing them to unfold and flip from their usual conformation to the scrapie shape. This change initiates a cascade in which newly converted molecules change the shape of other normal PrP molecules, and so on.

The collected studies argue persuasively that the prion is an entirely new class of infectious pathogen and that prion diseases result from aberrations of protein conformation. Whether changes in protein shape are responsible for common neurodegenerative diseases, such as Alzheimer's, remains unknown, but it is a possibility that should not be ignored.

Vocabulary

PrP (prion protein) プリオンタンパク質

conformational コンフォーメーションに関する

unfold たたんだものを広げる
flip 位置をひっくり返す
conformation コンフォーメーション
▶ Technical Terms

aberration 異常

neurodegenerative diseases 神経変性疾患

Technical Terms

コンフォーメーション(**conformation**)　分子の3次元的立体構造のこと。「立体配座」ともいう。同じ有機分子でも，それぞれの原子が占める空間的な位置は複数通りありうる場合がある。タンパク質や核酸などの生体分子では，分子の各部分のコンフォーメーションの違いによって分子全体の立体構造が大きく異なる。

私は15年前，ある種の中枢神経変性疾患を動物やまれには人間に引き起こす伝染性の病原体が，タンパク質だけでできていてほかには何も含んでいないと発表し，かなりの物議をかもした。当時，この考えは異端だった。伝染性の病気の運び手は宿主に感染を確立するために核酸（DNA や RNA）でできた遺伝物質を必要とするというのが定説だったからだ。最も単純な微生物であるウイルスでさえ，生存と複製に必要なタンパク質の合成の指令を，そうした遺伝物質に頼っている。また後に私たちが，これらの「タンパク質性感染粒子」，つまり私が「プリオン」と呼んだ病原物質が，伝染病だけでなく遺伝病の原因にもなっていると指摘したときにも，多くの科学者はやはり怪しいと考えた。そんな二面性を持つ物質は当時の医学界には知られていなかった。さらに，私たちがプリオンは正常タンパク質分子の形を変化させて危険なタンパク質分子に変えるという信じ難い方法で増殖すると結論づけたときにも，反発に見舞われた。しかし現在では，多くの実験的・臨床的データによって，3つすべての点で私たちが正しかったことが証明された。

現在知られているプリオン病はすべて致死性であり，海綿状脳症と呼ばれることもある。脳が穴だらけになる場合が多いため，この名がある。プリオン病は発病までに何年もかかり（人間では何十年もかかる場合もある），動物に広く見られる。最も有名なのはヒツジやヤギに見られる「スクレイピー」だ。ウシの海綿状脳症は狂牛病とも呼ばれ，最も心配されている（ヒトのプリオン病には認知症を引き起こすクロイツフェルト・ヤコブ病などがある）。

私たちはプリオンタンパク質（PrP）が核酸の助けなしに複製して病気を引き起こすことを示したほか，スクレイピー PrP がどのように細胞に伝染していくのかをつかんだ。詳細の多くは未解明だが，一点だけは極めて明白で，正常 PrP とスクレイピー PrP の大きな違いはタンパク質のコンフォーメーションである。明らかに，スクレイピータンパク質は正常なPrP分子に接触し，何らかの方法によってその普通のコンフォーメーションをスクレイピー型に変えることによって，自分を増やしていく。この変化が連鎖的に続いて，正常 PrP からスクレイピー型への変化がどんどん増えていくのだ。

連の研究によって，プリオンがまったく新しいタイプの感染性病原体であり，プリオン病がタンパク質のコンフォーメーションの異常によって生じることが説得力をもって示された。こうしたタンパク質の形の変化が，アルツハイマー病など一般的な神経変性疾患に関係しているかどうかはまだ不明だが，その可能性を無視すべきではない。

The Molecular Logic of Smell

匂いの分子生物学（1995年掲載）

R. アクセル（2004年受賞）

リチャード・アクセル（Richard Axel, 1946年～）は米国の神経科学者。2004年，「嗅覚受容体および嗅覚系組織の発見」よって，教え子の生物学者リンダ・バック（Linda B. Buck, 1947年～）とともにノーベル生理学・医学賞を受賞した。

アクセルとバックは嗅覚の受容器がGタンパク質共役受容体の一種であることを具体的に示して1991年に論文発表した。また，哺乳動物のDNAを解析し，嗅覚受容体をコードしている約1000の遺伝子を突き止めた。20世紀後半に急進展した遺伝子解析と分子生物学的な手法によって，匂いの知覚のメカニズムに迫った成果だ。

ノーベル賞受賞の10年ほど前にアクセルが執筆した以下の記事は，バックとともに取り組んだこの研究の興奮を伝えている。

初出：SCIENTIFIC AMERICAN October 1995, 日経サイエンス 1995年12月号
抜粋掲載：SCIENTIFIC AMERICAN July 2014, 日経サイエンス 2014年11月号

The basic anatomy of the nose and olfactory system has been understood for some time. In mammals, for example, the initial detection of odors takes place at the posterior of the nose, in the small region known as the olfactory epithelium. A scanning electron micrograph of the area reveals two interesting types of cells. In this region, millions of neurons, the signaling cells of sensory systems, provide a direct physical connection between the external world and the brain. From one end of each neuron, hairlike sensors called cilia extend outward and are in direct contact with the air. At the other end of the cell, a fiber known as an axon runs into the brain. In addition, the olfactory epithelium contains neuronal stem cells, which generate olfactory neurons throughout the life of the organism. Unlike most neurons, which die and are never replaced, the olfactory sensory neurons are continually regenerated.

When an animal inhales odorous molecules, these structures bind to specialized proteins, known as receptor proteins, that extend from the cilia. The binding of odors to these receptors initiates an electrical signal that travels along the axons to the olfactory bulb, which is located in the front of the brain, right behind the nose itself. The olfactory bulb serves as the first relay station for processing olfactory information in the brain; the bulb connects the nose with the olfactory cortex, which then projects to higher sensory centers in the cerebral cortex, the area of the brain that controls thoughts and behaviors. Somewhere in this arrangement lies an intricate logic that the brain uses to identify the odor detected in the nose,

Vocabulary

olfactory system 嗅覚系
mammal 哺乳動物, 哺乳類
olfactory epithelium 嗅上皮
neuron ニューロン
cilia 繊毛
axon 軸索
neuronal stem cell 神経幹細胞
sensory neuron 感覚ニューロン
receptor protein 受容体タンパク質 ▶ Technical Terms
olfactory bulb 嗅球
olfactory cortex 嗅皮質
project 投射する
sensory center 感覚中枢
cerebral cortex 大脳皮質

Technical Terms

受容体タンパク質(**receptor protein**)　受容体を構成しているタンパク質。ここでいう受容体は, 細胞に存在して外部の物質などと選択的に結びついてシグナルとして受け取る物質のことで, 一般的には細胞膜上にあるタンパク質のこと。受容体の働きを阻害したり促進したりすると生化学反応をコントロールできるので, 受容体に作用して望ましい効果を生み出す疾病治療薬の研究開発が大きな流れになっている。

distinguish it from others, and trigger an emotional or behavioral response.

To probe the organization of the brain, my co-workers and I began where an odor is first physically perceived—at the odor receptor proteins. Instead of examining odor receptors directly, Linda Buck, then a postdoctoral fellow in my laboratory, and I set out to find the genes encoding odor receptors. Genes provide the template for proteins, the molecules that carry out the functions of cells.

Using the technique of gene cloning, we were able to isolate the genes encoding the odor receptors. This family of receptor genes exhibited several properties that suited it to its role in odor recognition. First, the genes encoded proteins that fall squarely within a previously described group of receptors that pass through the cell membrane of the neuron seven times; these receptors activate signaling proteins known as G proteins. Early studies by Doron Lancet of the Weizmann Institute of Science and Randall R. Reed of the Johns Hopkins School of Medicine have established that odor receptors, too, use G proteins to initiate the cascade of events resulting in the transmission of an electrical impulse along the olfactory sensory axon. Second, the genes encoding the odor receptor proteins are active only in olfactory neurons. Although nearly every cell of the body carries a copy of every gene, many genes are expressed only in specialized cells.

Finally, a broad range of odor receptor genes appears to mirror the striking range of odors. By examining DNA from a variety of mammals, including humans, we

Vocabulary

gene cloning 遺伝子クローニング

fall squarely within~ きちんと〜に含まれる

G proteins Gタンパク質
▶ Technical Terms

initiate 開始する

axon 軸索

Technical Terms

Gタンパク質(**G proteins**) ホルモンや神経伝達物質が細胞膜上の受容体に結合して細胞内に一連の生化学反応を起こすとき,その信号伝達を助けているタンパク質。グアニンヌクレオチド結合タンパク質の略。

determined that around 1,000 genes encode 1,000 different odor receptors. (Each type of receptor is expressed in thousands of neurons.) Given that mammalian DNA probably contains around 100,000 genes, this finding indicates that 1 percent of all our genes are devoted to the detection of odors, making this the largest gene family thus far identified in mammals. The enormous amount of genetic information devoted to smell perhaps reflects the significance of this sensory system for the survival and reproduction of most mammalian species.

Vocabulary

devote to 振り向ける

鼻と嗅覚系の基本的な解剖学はしばらく前からわかっている。例えば哺乳類では，最初の匂い検出は鼻の中の後方にある「嗅上皮」という小さな領域で行われる。走査型電子顕微鏡でこの部分を観察すると，2種類の興味深い細胞が見られる。嗅上皮には感覚系の情報を伝えるニューロンが数百万個並び，外界と脳を直接物理的につないでいる。各ニューロンの一端から繊毛と呼ばれる毛髪状の感覚器が伸び，外気に接している。細胞の他端からは，軸索という線維が脳へ向かっている。さらに嗅上皮には神経幹細胞があり，一生を通じてニューロンを生み出している。死ぬとそれきりである他の多くのニューロンとは異なり，嗅覚系の感覚ニューロンは絶えず再生されている。

動物が匂いの分子を吸い込むと，繊毛の上に広がる特殊化した受容体タンパク質にその分子が結合する。匂い分子と受容体の結合によって電気信号が生じ，これが軸索に沿って伝わって，鼻の真後ろで脳の前面にある「嗅球」に達する。嗅球は脳で嗅覚情報を処理する第1中継局であり，鼻と脳の嗅皮質をつないでいる。嗅皮質は，思考と行動をコントロールしている大脳皮質にあるより高次の感覚中枢へ情報を投射する。この一連のつながりのどこかに，鼻で検出された匂いを同定し，他の匂いと区別し，情動や行動の応答を引き起こすのに脳が用いている複雑な論理が存在している。

脳のこの機構を探るため，私は共同研究者とともに，匂いが最初に身体的に感知される場所から調べ始めた。匂い受容体タンパク質だ。私はポスドク研究員のバック（Linda Buck）とともに，匂い受容体を直接調べる代わりに，匂い受容体をコードしている遺伝子を探す研究を始めた。遺伝子は細胞内で機能を実行するタンパク質のもととなる“ひな型”だ。

遺伝子クローニングの技術を用いて，私たちは匂い受容体をコードしている遺伝子群を単離することができた。この遺伝子ファミリーは匂い感知という役割にふさわしい特性をいくつか示していた。第1の特性は，これらの遺伝子がコードしているタンパク質が，既知の一群の受容体と同じグループに分類されることだ。これらの受容体はニューロンの細胞膜を7回貫通し，Gタンパク質というシグナルタンパク質を活性化させる。ワイツマン科学研究所のランセット（Doron Lancet）とジョンズ・ホプキンズ大学医学部のリード（Randall R. Reed）

らの初期の研究により，匂い受容体もGタンパク質を使って，嗅覚路の軸索に沿う電気パルスの伝達に至る一連の反応を始動することがわかっている。2番目の特徴は，匂い受容体タンパク質をコードする遺伝子は嗅覚ニューロンだけで発現している点だ。身体の細胞はほぼすべて，全遺伝子のコピーを持っているが，多くの遺伝子は特定の細胞においてのみ発現する。

そして最後に，匂い受容体遺伝子の多様さは，匂いの驚くほどの多様さを反映していると思われる。私たちはヒトを含む多くの哺乳動物のDNAを調べ，約1000個の遺伝子が1000種類の異なる匂い受容体をコードしていることを突き止めた（各タイプの受容体がそれぞれ数千個のニューロンで発現している）。哺乳動物のDNAが含む遺伝子がおそらく約10万個であることを考えると，この発見はヒトの全遺伝子の1%が匂い検出に充てられており，哺乳動物でこれまでに同定された遺伝子の中で最大のファミリーであることを示している。このように膨大な量の遺伝情報が嗅覚に充てられているのは，多くの哺乳動物の生存と繁殖にとって嗅覚という感覚系が重要であることの反映であろう。

Telomeres, Telomerase and Cancer

テロメアとがん（1996年掲載）

C. W. グライダー／E. H. ブラックバーン（ともに2009年受賞）

キャロライン・ウィドニー・グライダー（Carolyn Widney Greider, 1961年～）とエリザベス・ヘレン・ブラックバーン（Elizabeth Helen Blackburn, 1948年～）はいずれも米国の分子生物学者で，グライダーはブラックバーンの教え子。2人は2009年,「テロメアとテロメラーゼ酵素が染色体を保護する機序の発見」によって，米国の分子生物学者ジャック・ショスタク（Jack W. Szostak, 1952年～）とともにノーベル生理学・医学賞を受賞した。

テロメアとは染色体の両端にある構造で，染色体の本体を保護する“キャップ”の役割を果たしている。グライダーとブラックバーンはテトラヒメナという繊毛虫からテロメア配列を同定し，テロメアを伸長するテロメラーゼという酵素を1984年に発見した。

細胞の分裂・増殖でDNAが複製される際，テロメア部分は完全には複製されず，しだいに短くなっていく。短くなりすぎると細胞は分裂しても存続できず，死滅すると考えられている。

ノーベル賞受賞の13年前に書かれた以下の記事では，テロメアの一般的解説に加え，テロメラーゼを阻害することによってがん細胞を死に追い込む方法が提案されている。実際，テロメラーゼを標的とする抗がん剤の開発が進んでいるところだ。

初出：SCIENTIFIC AMERICAN February 1996, 日経サイエンス1996年4月号
抜粋掲載：SCIENTIFIC AMERICAN June 2011, 日経サイエンス2011年11月号

During the past 15 years, investigations have led to identification of an extraordinary enzyme named telomerase that acts on telomeres [the tips of chromosomes] and is thought to be required for the maintenance of many human cancers. Cancers arise when a cell acquires multiple genetic mutations that together cause the cell to escape from normal controls on replication and migration. As the cell and its offspring multiply uncontrollably, they can invade and damage nearby tissue. Some parts may break away and travel to parts of the body where they do not belong, establishing new malignancies at distant sites.

The notion that telomerase might be important to the maintenance of human cancers was discussed as early as 1990. But the evidence did not become compelling until recently. Findings have led to an attractive but still hypothetical model for the normal and malignant activation of telomerase by the human body. According to this model, telomerase is made routinely by cells of the germ line in the developing embryo. Once the body is fully formed, however, telomerase is repressed in many somatic [nongerm] cells, and telomeres shorten as cells reproduce. When telomeres decline to a threshold level, a signal is emitted that prevents the cells from dividing further.

If, however, cancer-promoting genetic mutations block issuance of such safety signals or allow cells to ignore them, cells will continue to divide. They will also presumably continue to lose telomeric sequences and undergo chromosomal alterations that allow further, possibly carcinogenic mutations to arise. When

Vocabulary

enzyme 酵素
▶ Technical Terms
telomerase テロメラーゼ
telomere テロメア
▶ Technical Terms
chromosome 染色体
mutation 変異
replication 増殖
migration 移動
malignancy 悪性組織，腫瘍
compelling 説得力のある
germ line 生殖細胞系列
embryo 胚，胎児
repress 抑制する
somatic cell 体細胞
issuance 発行，発信
carcinogenic 発がん性の

Technical Terms

酵素（**enzyme**）　生化学反応を促進する触媒として働くタンパク質。生体内の化学反応のほとんどに酵素が関与しており，様々な酵素が知られている。
テロメア（**telomere**）　染色体の末端部をなす構造で，“保護キャップ”にたとえられる。細胞が分裂して DNA が複製されるとき，この末端部分は完全には複製されず，細胞分裂のたびに少しずつ短くなっていく。

telomeres are completely or almost completely lost, cells may reach a point at which they crash and die. But if the genetic derangements of the pre-crisis period lead to the manufacture of telomerase, cells will not fully lose their telomeres. The shortened telomeres will be rescued and maintained. In this way, the genetically disturbed cells will gain the immortality characteristic of cancer.

This scenario has generally been borne out by the evidence, although some advanced tumors lack telomerase, and some somatic cells—notably the white blood cells known as macrophages and lymphocytes—have recently been found to make the enzyme. Nevertheless, on balance, the collected evidence suggests that many tumor cells require telomerase in order to divide indefinitely.

The presence of telomerase in various human cancers and its absence in many normal cells mean the enzyme might serve as a good target for anticancer drugs. Agents able to hobble telomerase might kill tumor cells (by allowing telomeres to shrink and disappear) without disrupting the functioning of many normal cells. In contrast, most existing anticancer therapies disturb normal cells as well as malignant ones, and so are often quite toxic. Further, because telomerase occurs in numerous cancers, such agents might work against a broad array of tumors.

Vocabulary

derangement 混乱

immortality 不死化能, 不死性

bear out 支持する。borne は bear の過去分詞
white blood cell 白血球
▶ Technical Terms
macrophage マクロファージ
▶ Technical Terms
lymphocyte リンパ球
▶ Technical Terms

hobble 妨げる

Technical Terms

白血球(**white blood cell, leukocyte**)　もともとは赤血球の対語で, 血液中に見られる呼吸色素なしの細胞のことだが, 現在では免疫防御を担当している細胞(顆粒球, リンパ球, 単球, マクロファージなど)を総称としてこう呼ぶ場合が多い。
マクロファージ(**macrophage**)　白血球の一種。体内をアメーバのように遊走して, 傷ついた細胞や外部から侵入した細菌などの異物を貪食している。自分の表面に抗原を提示してT細胞(リンパ球の一種)を活性化するなど, 免疫機能で重要な役割を担っている。
リンパ球(**lymphocyte**)　免疫機能を担っている白血球で, T細胞, B細胞, ナチュラルキラー細胞などの種類がある。

過去15年間に，テロメア（染色体の末端部分）に特異的に作用する酵素テロメラーゼが発見され，この酵素がヒトの多くのがんを維持するのに必須と考えられるようになった。がんは細胞が多くの遺伝子変異を蓄積した結果，増殖と体内移動に関する正常な制御から逸脱して起こる。この細胞とその子孫細胞は無制限に増殖し，周囲に浸潤して周辺組織を害する。一部はもとの組織を脱して遠く離れた別の組織まで移動し，そこに新しいがん組織を形成する。

ヒトのがんの維持にテロメラーゼが重要な役割を果たしているらしいという考え方は，早くも1990年に議論されていた。しかし確定的な証拠が得られたのはごく最近のことだ。複数の発見をもとに，人体によるテロメラーゼの正常な，そして異常な活性化についての，まだ仮説の段階ではあるが魅力的なモデルが導かれた。このモデルによると，発生途中の胎児では生殖細胞系列によってテロメラーゼがごく普通に作られている。しかし胎児の身体が完全に形成すると，多くの体細胞（生殖細胞系列以外の細胞）でテロメラーゼは抑制され，体細胞が分裂するたびにテロメアは短くなる。テロメアがある閾値まで短くなると，以降の細胞分裂を妨げるシグナルが発せられる。

しかし，がん化を促す遺伝子変異が起こると，こうしたシグナルの発信が阻害されるか，発信シグナルが無視されるようになり，細胞は分裂・増殖を続ける。この結果，テロメアの短縮がますます進むだろうし，染色体の変化が起こって，がん化につながる変異がさらに蓄積するだろう。テロメアが完全に失われるか，ほとんどなくなると，その細胞は破綻して死ぬ点に至るだろう。しかし，それ以前に生じた遺伝子の変化によってテロメラーゼが作られている場合，その細胞ではテロメアが完全に失われることはなく，短縮したテロメアが救われ維持されるだろう。こうして，遺伝的変化を来したこの細胞は，がんの特徴である不死化能を獲得することになる。

このシナリオを支える証拠が一般に得られている。ただし，一部の進行性腫瘍はテロメラーゼを欠いているし，ある種の正常な体細胞（特にマクロファージやリンパ球などの白血球）がテロメラーゼを作っていることが最近になってわかった。とはいえ，すべてを考慮すると，これまでの研究結果は多くのがん細胞が果てしなく分裂・増殖を続けるにはテロメラーゼが必要であることを示唆している。

多くの正常組織にはテロメラーゼがなく，さまざまながん組織にテロメラーゼが存在するという事実は，テロメラーゼが抗がん剤のターゲットとして適している可能性を意味している。テロメラーゼを阻害する薬物は，がん細胞を殺す（テロメアの短縮・消失を許すことによって）一方で，多くの正常細胞の機能を混乱させずにすむだろう。これと対照的に，既存の抗がん剤の大半は，がん細胞と同時に正常細胞も混乱させるので，重い副作用が生じる場合が多い。さらに，テロメラーゼはほとんどのがんで生じているので，テロメラーゼに作用する抗がん剤は広範囲のがんに効果があるだろう。

The Birth of Complex Cells

真核細胞はどのように生まれたか（1996年掲載）

C. ド・デューブ（1974年受賞）

クリスチャン・ルネ・ド・デューブ（Christian René de Duve，1917～2013年）はベルギーの細胞生物学者・生化学者。1974年，「細胞の構造的機能的組織に関する発見」によって，ベルギーの細胞生物学者アルベルト・クラウデ（Albert Claude，1899～1983年）および米国の細胞生物学者ジョージ・エミール・パラーデ（George Emil Palade，1912～2008年）とともにノーベル生理学・医学賞を受賞した。

ド・デューブはペルオキシソームやリソソームなどの細胞小器官を発見し，この業績がノーベル賞の授賞理由になった。そして，この研究がもととなって，現在の多細胞生物（真核生物）の細胞に存在するミトコンドリアなどの細胞小器官が，もともとは細菌などの原核生物が入り込んだ名残であるという「細胞内共生説」が定着して現在に至っている。

以下の記事はノーベル賞受賞から20年あまり後に執筆されもので，細胞内共生説の考え方を説得力豊かに解説している。

初出：SCIENTIFIC AMERICAN April 1996, 日経サイエンス 1996年6月号
抜粋掲載：SCIENTIFIC AMERICAN June 2011, 日経サイエンス 2011年11月号

About 3.7 billion years ago the first living organisms appeared on the earth. They were small, single-celled microbes not very different from some present-day bacteria. Prokaryotes turned out to be enormously successful. Thanks to their remarkable ability to evolve and adapt, they spawned a wide variety of species and invaded every habitat the world had to offer. The living mantle of our planet would still be made exclusively of prokaryotes but for an extraordinary development that gave rise to a very different kind of cell, called a eukaryote because it possesses a true nucleus. Today all multicellular organisms consist of eukaryotic cells. Eukaryotic cells most likely evolved from prokaryotic ancestors. But how?

Appreciation of this astonishing evolutionary journey requires a basic understanding of how the two fundamental cell types differ. Eukaryotic cells are much larger than prokaryotes (typically some 10,000 times in volume). In prokaryotes the entire genetic archive consists of a single chromosome made of a circular string of DNA that is in direct contact with the rest of the cell. In eukaryotes most DNA is contained in more highly structured chromosomes that are grouped within a well-defined central enclosure, the nucleus. Most eukaryotic cells further distinguish themselves from prokaryotes by having in their cytoplasm up to several thousand specialized structures, or organelles, about the size of a prokaryotic cell. The most important of such organelles are peroxisomes (which serve assorted metabolic functions), mitochondria (the power factories of cells) and, in algae and plant cells, plastids (the sites of photosynthesis).

Vocabulary

bacteria 細菌，バクテリア
prokaryote 原核生物
▶ Technical Terms
spawn 生む，引き起こす
habitat 生息環境
mantle 包み込むもの，外層
eukaryote 真核生物
▶ Technical Terms
nucleus （細胞の）核
multicellular organism 多細胞生物
chromosome 染色体
cytoplasm 細胞質
organelle 細胞小器官
▶ Technical Terms
peroxisome ペルオキシソーム
mitochondria ミトコンドリア
algae 藻類
plastid 色素体
photosynthesis 光合成

Technical Terms

原核生物（**prokaryote**）　核膜で覆われた明確な細胞核を持たない細胞（原核細胞）からなる生物。細菌や古細菌など。
真核生物（**eukaryote**）　細胞核を持つ真核細胞からなる生物。単細胞生物の一部と，すべての多細胞生物がこれにあたる。
細胞小器官，オルガネラ（**organelle**）　細胞質のなかにあって一定の機能を果たしている構造体の総称。例えばミトコンドリア，小胞体，ゴルジ体など。

Biologists have long suspected that mitochondria and plastids descend from bacteria that were adopted by some ancestral host cell as endosymbionts (a word derived from Greek roots that means "living together inside"). The most convincing evidence is the presence within these organelles of a vestigial—but still functional—genetic system. That system includes DNA-based genes, the means to replicate this DNA, and all the molecular tools needed to construct protein molecules from their DNA-encoded blueprints. Endosymbiont adoption is often presented as resulting from some kind of encounter—aggressive predation, peaceful invasion, mutually beneficial association or merger—between two typical prokaryotes. There is a more straightforward explanation—namely, that endosymbionts were originally taken up in the course of feeding by an unusually large host cell that had already acquired many properties now associated with eukaryotic cells. Many modern eukaryotic cells—white blood cells, for example—entrap prokaryotes. On a rare occasion, both captor and victim survive in a state of mutual tolerance that can turn into mutual assistance and, eventually, dependency. Mitochondria and plastids thus may have been a host cell's permanent guests.

Vocabulary

adopt 採用する, 受け入れる
endosymbiont 内部共生したもの

vestigial 痕跡の, 名残の

predation 捕食

white blood cell 白血球
entrap とらえて閉じ込める
captor とらえる者

dependency 依存

約37億年前，最初の生物が地球上に現れた。小さな単細胞の微生物で，現在の細菌とたいして変わらない。これらの原核生物は大きな成功を収めた。進化し適応する優れた能力のおかげで，さまざまな種を生じ，世界の隅々にまですみついた。もし，まったく異なるタイプの細胞を持つ真核生物（真の細胞核があるのでこう呼ばれる）を生む異例の発展がなかったら，地球を包む生物の層はいまも原核生物だけでできていたことだろう。現在，すべての多細胞生物は真核細胞からなっている。真核細胞は古くからの原核細胞から進化したと考えるのが自然だろう。だが，どのようにして？

この驚くべき進化の旅を正しく理解するには，これら2つの基本的な細胞タイプがどう異なっているかについて基礎的な理解が必要だ。真核細胞は原核細胞よりもはるかに大きい（一般に体積は約1万倍）。原核細胞ではすべての遺伝情報が環状のDNAでできた単一の染色体に収められ，その染色体が細胞の他の部分とじかに接している。これに対し真核細胞では，DNAの大部分がより高度に構造化された染色体に含まれていて，それらの染色体が細胞核という輪郭のはっきりした中央部の囲いのなかに収められている。このほか大部分の真核細胞は細胞質のなかに数千個にのぼる特殊化した構造体，オルガネラ（細胞内小器官）を持っている点でも原核細胞と異なる。オルガネラはそれぞれが原核細胞と同じくらいの大きさであり，最も重要なものにペルオキシソーム（さまざまな代謝機能を発揮する）やミトコンドリア（細胞のエネルギー工場）のほか，藻類と植物の細胞には色素体（光合成の場）がある。

ミトコンドリアと色素体は，何らかの祖先細胞が内部共生体として取り込んだ細菌がもとになっているのではないかと，生物学者はかねて疑ってきた。最も説得力のある証拠は，これらのオルガネラのなかに，痕跡的な，しかしまだ機能を発揮する遺伝システムが存在していることだ。そのシステムにはDNAを基本とする遺伝子と，そのDNAを複製する手段，DNAに描かれた青写真からタンパク質分子を組み立てるのに必要とされるすべての分子装置が含まれている。内部共生による取り込みは，2種の典型的な原核生物の遭遇（攻撃的な捕食，平和裡の侵入，相互に有利となる連携あるいは合体）の結果だとされることが多い。しかし，もっと明快な説明がある。つまり，真核細胞に見られるのと同じ多くの特性をすでに獲得ずみの異常に大きな細胞があって，それが他の原核生物を食べて取り込んだものが，そもそもの細胞内共生体であるという見方だ。現代の真核細胞の多く，例えば白血球などは，原核細胞を捕らえて吸収する。まれには，捕らえた側と犠牲者が相互に相手を許容する状態に入って生き延び，それが相互助力，さらには相互依存に進むことがある。ミトコンドリアと色素体は，このようにして宿主細胞に永遠に腰を落ち着ける客となったのかもしれない。

物理学賞
素粒子から宇宙まで

光とは何か What Is Light?

物質の原子構造を探るX線の指 X-Ray Fingers Feel Out the Atomic Structure of Matter

宇宙線に秘められたメッセージ The Secret Message of the Cosmic Ray

原子核の構造 The Structure of the Nucleus

反陽子 The Antiproton

光メーザー Optical Masers

X線星 X-ray Stars

超新星爆発のメカニズム How a Supernova Explodes

ヒッグス・ボソンは実在するか The Higgs Boson

究極の時間測定技術 Accurate Measurement of Time

宇宙の中の生命 Life in the Universe

ニュートリノの質量の発見 Detecting Massive Neutrinos

What Is Light?

光とは何か（1928年掲載）

E. O. ローレンス（1939年受賞）／ J. W. ビームズ

アーネスト・オーランド・ローレンス（Ernest Orlando Lawrence，1901～1958年）は米国の物理学者。「サイクロトロンの発明・開発およびその成果，特に人工放射性元素」によって，1939年のノーベル物理学賞を単独受賞した。

サイクロトロンは20世紀前半に標準的に使われた粒子加速器で，これを用いて多くの放射性元素が作られた。ウラン（原子番号92）よりも重い超ウラン元素のうち1950年代までに発見された大半は，ローレンスが所長を務めていたバークレー放射線研究所（現在の米国立ローレンス・バークレー研究所）で合成された。また，103番目の元素ローレンシウムは彼の名にちなんでいる。

以下の記事はローレンスがサイクロトロンを発明する1931年以前に執筆された。加速器について述べた記事ではないが，20世紀初頭に進展した量子論によって光の本質のとらえ方が大きく変わったことを，まだ20代だった気鋭の物理学者ローレンスが生き生きと伝えており興味深い。

共著者のビームズ（Jesse Wakefield Beams，1898～1977年）は米国の物理学者。執筆当時はエール大学でローレンスとともに光電効果の実験をしていた。後にバージニア大学教授。

初出：SCIENTIFIC AMERICAN April 1928
抜粋掲載：SCIENTIFIC AMERICAN July 2012, 日経サイエンス 2012年11月号

Light is one of the most familiar physical realities. All of us are acquainted with a large number of its properties, while some of us who are physicists know a great many more marvelous characteristics which it displays. The sum total of our knowledge of the physical effects produced by light is very considerable, and yet we have no satisfactory conception of what it is.

More than two centuries ago Newton conceived that light was corpuscular in nature; he believed that light consisted of little darts shooting through space. Others regarded light as a wave phenomenon; in a manner analogous to the propagation of waves in water, light waves were propagated in a medium pervading all space, called the ether. A lively controversy ensued between the adherents of these two conceptions of the nature of light, and as new experiments were carried out revealing more of its properties, it appeared that the undulatory theory accounted for many things quite unintelligible on the corpuscular hypothesis.

As time has progressed, many additional phenomena concerned with the interaction of light and matter have been discovered which are impossible of understanding on the wave theory and which have compelled scientists to revert to the conception of light which was in Newton's mind centuries ago. Such recent facts of observation suggest that light beams contain amounts of energy which are exact multiples of a definite smallest amount—a light quantum—just as matter seems to be made up of definite multiples of a smallest particle of matter or electricity—the electron. Thus, we have atomicity of light as well as atomicity of matter and electricity.

A seemingly very peculiar circumstance exists in this modern quantum theory of light, for the very thing concerned in the theory is entirely obscure.

Vocabulary

be acquainted with~ ～をよく知っている
considerable かなりの
corpuscular 粒子の
medium 媒質
pervade 広がる, 浸透する
ether エーテル
adherent 支持者, 信奉者
undulatory theory 波動説
unintelligible 不可解な
hypothesis 仮説, 説
interaction 相互作用
compel 強いる
revert to~ ～に立ち返る
light quantum 光量子
atomicity 原子性
peculiar 奇妙な
obscure はっきりしない, 曖昧な

And so the question of the physical nature of quanta presents itself. Are they a yard or a mile or an inch in length, or are they of infinitesimal dimensions? Many experimental facts can be interpreted as indicating that quanta are at least a yard in length, yet nothing really certain can be inferred from past observations. The dimensions in space of the quanta remain complete mysteries.

There is at least one way of measuring the length of quanta, provided that the scheme may be carried out in practice, which is essentially as follows: Suppose one had a light shutter that could obstruct or let pass a beam of light as quickly as desired. Such an apparatus would be able to cut up a beam of light into segments, much in the same way that a meat cutter slices a bologna sausage. It is clear that if the slices of the light beam so produced were shorter than the light quanta in the beam, the short light flashes coming from the shutter would contain only parts of quanta. In effect, the apparatus would be cutting off the heads or tails of quanta. To eject an electron from a metal surface a whole quantum is necessary because part of one quantum does not contain enough energy to do the trick. One therefore would definitely establish an upper limit to the length of light quanta by simply observing the shortest light flashes able to produce a photo-electric effect.

One does not have to be very familiar with mechanical things to realize that no mechanical shutter could possibly work at this speed. Happily, however, Nature has endowed matter with properties other than purely mechanical ones. By making use of a certain electro-optical

Vocabulary

quantum 量子(ここでは光量子)。quanta は複数形
infinitesimal 無限小
infer 推測する
obstruct 邪魔する，遮る
apparatus 装置
segment 断片
eject 放出させる
photo-electric effect 光電効果
▶ Technical Terms
endow 付与する，与える

Technical Terms

光電効果(**photo-electric effect**) 物質が光を吸収して外部に自由電子を放出あるいは伝導電子(固体中を移動して電流を担う電子)を生じる現象のこと。光が波長に応じたエネルギーを持つ粒子であると考えることによってこの現象をあざやかに説明したアインシュタイン(Albert Einstein)の 1905 年の理論は量子論の基礎となり，1921 年のノーベル物理学賞を受賞した。

property of some liquids a device was conceived which actually operated as a shutter, turning on and off in about one ten thousand millionth of a second.

The short flashes of light produced in this way were allowed to fall on a sensitive photo-electric cell, and it was found that the cell responded to the shortest flashes obtained—which were only a few feet in length.

The importance of this simple experimental observation cannot be overestimated, for it definitely demonstrated that light quanta are less than a few feet in length and probably occupy only very minute regions of space.

Vocabulary

photo-electric cell 光電素子

overestimate 過大評価する

光は物理的実体のなかでも最もなじみ深い。誰でも光の性質について多くのことをご存じだし，私たち物理学者は光が示すさらに多くの驚くべき特徴を知っている。光がもたらす物理的効果に関する私たちの知識量は相当なものだが，そもそも光とは何かという点については，私たちは十分な概念を持ち合わせていない。

200年以上前，ニュートンは光が本質的に粒子であると考えた。光は空間のなかを飛ぶ小さな投げ矢でできていると考えたのだ。他の人たちは光を波動現象とみなした。水の中を波が伝わるのと同様に，光波は空間を満たしているエーテルという媒質を伝わると考えた。光の本質について，これら2つの考え方の間で活発な論争が続いたが，新たな実験によって光の性質についてより多くのことが明らかになるにつれ，粒子説ではまったく不可解な多くの事柄を波動説なら説明できるように思われた。

さらに時がたつと，光と物質の相互作用に関する多数の新現象が発見された。それらは波動説では理解不能であり，科学者たちはニュートンの考え方に立ち戻ることを強いられた。そうした最近の観測事実は，光線の総エネルギーが明確な最小単位（光量子）のきっちり整数倍になっていることを示している。物質が物質と電気の最小単位である電子という粒子からできていると考えられるのと同様だ。こうして，私たちは物質と電気の原子性に加え，光の原子性を手にした。

この現代的な光の量子論をめぐる状況は非常に奇異に思える。それが論じている光というもの自体が，まるではっきりしていないのである。

光量子の物理的本質についても同じことがいえる。光量子の大きさは1mなのか1kmなのか1cmなのか，あるいは無限小なのか？　多くの実験事実は量子の長さが少なくとも1mはあることを示していると解釈できるが，これまでの観察からは確かなことは何も推定できない。光量子の空間的な大きさはいまだに完全な謎なのだ。

光量子の長さを測る方法が，実際に実行できればだが，少なくとも1つはある。その手順は以下のようなものだ。いくらでも素早く開閉できるシャッターがあって，それで光線を遮ったり通したりできるとしよう。この装置を使えば，肉切り包丁でボローニャソーセージを切るように，1つの光線を小さな断片に切り分けられるだろう。こうして作った断片がそのビームの光の量子よりも短くなったなら，シャッターからやってくるその短い閃光は明らかに光量子の一部分からできていることになる。光の量子の頭や尻尾の部分を切り落としているようなものだ。一方，金属に光を当てて表面から1個の電子を放出させるには，光量子の一部分ではエネルギーが不十分なため，量子まるまる1個が必要だ。したがって，光電効果を引き起こす最短の閃光を観測すれば，それが光の量子の長さの上限を与えることになる。

機械に特に詳しくない人でも，そんな高速で開閉する機械式シャッターなどありえないとわかるだろう。だが幸いなことに，自然が物質に与えた特性は純機械的なものだけではない。ある種の液体が持つ電気的・光学的特性を利用して，実際にシャッターとして機能する装置が考案された。約10億分の1秒で開閉する。

こうして作られた短い閃光を高感度の光電素子で受け，素子が長さ数十cmの最も短い閃光に反応することが確認された。

この単純な実験の重要性はいくら強調してもしすぎにはならない。光の量子が長さにして数十cmに満たず，おそらくは空間のごく微小な領域を占めるにすぎないことを明確に示したのだ。

X-Ray Fingers Feel Out the Atomic Structure of Matter

物質の原子構造を探るX線の指（1930年掲載）

W. H. ブラッグ（1915年受賞）

ウィリアム・ヘンリー・ブラッグ（William Henry Bragg, 1862～1942年）は英国の物理学者。1915年,「X線による結晶構造解析に関する研究」によって，息子のウィリアム・ローレンス・ブラッグ（William Lawrence Bragg, 1890～1971年）とともにノーベル物理学賞を受賞した。

X線を用いた結晶構造解析法は物質の構造をとらえる基本技術として20世紀の物性物理学研究に不可欠のツールとなったことであまりにも有名だ。また通常の固体物質だけでなく，後にはDNAやタンパク質など生体分子の構造解析にも用いられ，新たなX線光源の開発と相まって幅広い科学分野に大きく寄与し続けている。

ノーベル賞受賞の15年後に書かれた以下の記事は，この技術の原点を平易に解説している。当時の最先端技術に対する大きな期待を，ノーベル賞に輝いた泰斗の文章から感じることができる。この技術もブラッグの栄誉も，いまなお輝き続けている。

初出：SCIENTIFIC AMERICAN December 1930
抜粋翻訳掲載：http://www.nikkei-science.com/?p=27936，2012年9月

Man, having the power to forecast the result of overcoming difficulties and the wish to try to overcome them, has devised various ingenious methods to help him in his task. Taking first of all the difficulties that depend on the inadequacy of his vision he has invented the microscope which gives him the power of seeing details thousands of times too fine to be perceived by the naked eye.

But there is a point which the microscope can not pass. With its aid we perceive what is very small, but not the "very very" small. There are details of the structure of the living cell, essential features in the composition of metals, cotton, silk, rubber, paint, bone, nerve, and a thousand other things which are hidden even from the microscope, and must always remain so hidden because the failure does not lie with the skill of the optician but with the incapacity of light itself.

The nature of radiation is in many respects a mystery, but we know enough about it to understand that we may talk of it in many of its important aspects as waves in some medium which we call the ether. If the radiation falls on any object, it is turned aside and modified in various ways. When our eyes are directed towards the object, they take in the modified rays, and we have learned by long practice to know, from these modifications, the nature of the object that has made them. That is "seeing."

The central point of the process is the act of scattering and modification. Now waves have a certain wavelength, and common experience of such waves as may be seen, for example, on the surface of the sea tells us that an object which is very much smaller than the length of the wave has no appreciable effect upon it. In just the same way there may be objects which are so small that they can not affect a ray of light, and such objects are forever

Vocabulary

overcome 克服する
ingenious 巧妙な, 独創的な
inadequacy 不十分, 不足
feature 特徴
optician 光学機器製造業者
incapacity 無能, 無力
radiation 放射, 光
medium 媒質
ether エーテル
turn aside わきへそらす
scattering 散乱
modification 変調
wavelength 波長
appreciable はっきり感知できる

invisible in the ordinary sense. The length of the light wave which our eyes can perceive lies within a short range on either side of a fifty-thousandth of an inch.

The X rays break down the barrier for us, and admit us to this immense field in which we want to be. They do so by virtue of their character as light waves, 10,000 times or so smaller than visible waves, but of exactly the same nature.

If a substance is such that all the atoms which compose it are arranged on one and the same pattern, so that the straight rows run through from side to side, the substance is a single crystal; crystalline character meaning, simply, perfect arrangement. But most substances, and especially those we handle every day, such as the metals, must be described as masses of small separate crystals.

However we try to deform [a bar of multi-crystalline material], there are always some crystals which resist being deformed in that particular way. And the various crystals back each other up according to some principle which we do not fully understand. Thus the properties of the bar depend upon its crystalline character. It is only the X ray that can tell us the internal arrangement of the crystal.

The X rays are of short enough wavelength to be turned aside or scattered by the atoms, when longer light waves are not. A single atom can, however, do very little. Here is where the regularity of crystal arrangement comes in. The unit of pattern is repeated an enormous number of times even in a crystal just visible to the naked eye. Whatever one of these units does in the way of

Vocabulary

admit 入ることを認める

by virtue of~ 〜によって

single crystal 単結晶
▶ Technical Terms

deform 変形する
multi-crystalline 多結晶の

back 支える，支持する

regularity 規則性
come in 役割を担う，関与する

in the way of 〜について

Technical Terms

単結晶（**single crystal**） 試料のどの部分に注目しても結晶軸の向きが同一で，原子配列の繰り返しパターンが一貫している結晶のこと。これに対し，通常の固体の多くは小さな単結晶の粒が様々な方位を向いて集まったもので，これは多結晶（**polycrystal**）と呼ばれる。

scattering, all the others do in regular order. The combined amount is perceptible, and so the crystalline character is detected.

Of course this is an indirect way of examining the structure. We do not perceive the individual atoms; we discover only their arrangements. But the knowledge so gained can be combined with other knowledge that we already possess and we have actually found ourselves able to decipher the patterns of Nature to an extent we did not dream of a few years ago.

Vocabulary

detect 検出する

perceive 知覚する

decipher 解読する, 読み解く人間は困難を克服した先の結果を予想する力と，困難を克服しようとする望みを持っており，その試みを助けるために様々な巧妙な方法を考案してきた。視覚の不足に伴う困難を克服するために顕微鏡を発明したのが好例で，顕微鏡は肉眼では小さすぎて感知できないものを何千倍にも拡大して詳細を見る力を与えた。

だが，顕微鏡では超えられない限界がある。私たちは顕微鏡の助けでとても小さなものを知覚するが，“とてもとても小さなもの”は見られない。生きている細胞の詳細な構造，各種の金属や綿毛，絹糸，ゴム，塗料，骨，神経といったものを構成する成分の重要な特徴など，多くの事柄は顕微鏡でも隠されたままであり，今後も決して見ることはできない。それは光学機械屋の腕が悪いからではなく，光そのものが無力になるからだ。

放射の本質は多くの面で謎だが，光をエーテルという媒質のようなもののなかを伝わる波だと考えると，その重要な性質の多くを語れることはわかっている。放射が物体にぶつかると向きをそらされ，様々に変えられる。私たちの目がその物体に向いていれば，この変調された光が目に入る。そして私たちは，それら光の変調から，その変調を生んだ物体が何かを知るすべを長い経験を通じて身につけている。これが“見る”ということだ。

この過程の中心となっているのが散乱と変調の作用だ。波はある波長を持っており，誰でも海面の波などを見て知っているように，波長よりもずっと小さな物体は波にほとんど影響を及ぼさない。それと同様に，非常に小さな物体は光線に影響を及ぼすことができず，そうした物体は通常の意味では決して見ることができない。人間の目が感知できる光の波長は，幅が5万分の1インチの狭い範囲にある。

X線はこの限界を破って，私たちが望む素晴らしいミクロの領域に導いてくれる。それが可能なのは，X線が可視光の1万分の1の小さな光波であると同時に，光の波としての性質はまったく同じであるという本質による。

物質質を構成しているすべての原子が同一のパターンで配置していて，原子が並んだまっすぐな列が端から端まで通っている場合，その物質は単結晶といわれる。完璧な配列の結晶ということだ。だが，ほとんどの物質，特に金属など私たちが日常手にする物質は，小さな結晶が寄り集まったものとして記述しなければならない。

多結晶物質の棒をいかに変形しようとしても，一部にはそのような変形に抵抗する結晶が必ず存在する。そして，まだよくわかっていない何らかの原理によって，様々な結晶が互いを支え合っている。このように，棒の性質はその結晶の特徴によって決まるのだ。物体内部の結晶の配列を語ってくれるのはX線だけである。

通常の光波では波長が長すぎてだめだが，X線は原子によって散乱されるだけの十分に短い波長を持っている。しかし，1個の原子による散乱はごくわずかだ。そこで規則的な結晶配置の出番となる。肉眼でやっと見えるだけの小さな結晶でも，配列パターンの単位が膨大な回数にわたって繰り返している。そうした繰り返し単位の1つが散乱を起こしているとき，他のすべてのユニットも同じく規則的に散乱する。これらを足し合わせた結果は検知可能なものとなり，結晶の特徴が検出される。

もちろん，これは構造を調べる間接的な方法だ。個々の原子を知覚することはできない。その配列がわかるだけだ。しかし，こうして得た知識をすでに持っていた知識と組み合わせることが可能であり，自然界のパターンを数年前には想像もできなかったところまで読み解くことができるようになった。

The Secret Message of the Cosmic Ray

宇宙線に秘められたメッセージ（1933年掲載）

A. H. コンプトン（1927年受賞）

アーサー・コンプトン（Arthur Holly Compton，1892～1962年）は米国の実験物理学者。「コンプトン効果の発見」によって1927年のノーベル物理学賞を受賞した。ちなみに同年の物理学賞の半分は「霧箱（蒸気の凝縮により荷電粒子の飛跡を観察できるようにする方法）の考案」によってチャールズ・ウィルソン（Charles T. R. Wilson，1869～1959年）に授与された。

「コンプトン効果」とは物体にX線を照射すると波長の長いX線になって散乱される現象で，X線のエネルギーの一部が物質中の電子に移ることで起こる。こうした電磁波と荷電粒子の相互作用は自然界の様々なシーンで生じるが，その代表例といえるのが宇宙線（宇宙空間を飛び交っている荷電粒子）と電磁波の相互作用だ。

コンプトンも後に宇宙線の研究に力を注いだ。ノーベル賞受賞の6年後に執筆された以下の記事は当時の宇宙線研究の興奮を伝えている。抜粋記事の最後に宇宙線が持つ巨大なエネルギーに関する記述があるが，現在ではさらに桁違いに大きなエネルギーの宇宙線の存在がわかっている。高エネルギー宇宙線は超新星爆発やブラックホールにかかわる現象によって生まれると推定されているものの確証は得られておらず，現在も謎の解明に向けた努力が続いている。

初出：SCIENTIFIC AMERICAN July 1933
抜粋掲載：SCIENTIFIC AMERICAN July 2012，日経サイエンス 2012年11月号

The study of cosmic rays has been described as "unique in modern physics for the minuteness of the phenomena, the delicacy of the observations, the adventurous excursion of the observers, the subtlety of the analyses, and the grandeur of the inferences." These rays are bringing us, we believe, some important message. Perhaps they are telling us how our world has evolved, or perhaps news of the innermost structure of the atomic nucleus. We are now engaged in trying to decode this message.

About five years ago, two German physicists, Bothe and Kolhörster, did an experiment with counting tubes which convinced them that the cosmic rays are electrically charged particles. If this conclusion is correct, it means, however, that there should be a difference in intensity of the rays over different parts of the earth. For the earth acts as a huge magnet, and this huge magnet should deflect the electrified particles as they shoot toward the earth. The effect should be least near the magnetic poles, and greatest near the equator, resulting in an increasing intensity as we go from the equator toward the poles. A series of half a dozen different experiments designed to detect such effects resulted in inconclusive data.

Accordingly, with financial help from the Carnegie Institution, a group of us at the University of Chicago have organized nine different expeditions during the past 18 months, going into different portions of the globe to measure cosmic rays from sea level to the tops of mountains nearly four miles high in the Andes and the Himalayas. Two capable mountaineers, Carpe and Koven, lost their lives on a glacier on the side of mighty Mt. McKinley in Alaska, but they got the highest altitude data

Vocabulary

cosmic ray 宇宙線
▶ Technical Terms
excursion 遠征
grandeur 壮大さ, 雄大さ
atomic nucleus 原子核
counting tube 計数管
electrically charged particle 荷電粒子
intensity 強度
deflect そらす
magnetic pole 磁極
inconclusive 決定的でない, まちまちな
expedition 遠征調査
capable 有能な
altitude 標高

Technical Terms

宇宙線(**cosmic ray**)　宇宙空間を飛び交っている高エネルギーの粒子のこと。多くは陽子だが, ヘリウムの原子核(α粒子)やより重い元素の原子核もある。これらの一次宇宙線が地球大気にぶつかって生じた放射は二次宇宙線と呼ばれる。

yet obtained for latitudes so close to the pole.

On bringing together the results of these expeditions, it was found that the cosmic ray intensity near the poles is about 15 percent greater than near the equator. Furthermore, it varies with latitude, just as predicted, due to the effect of the earth's magnetism on incoming electrified particles. At high altitudes the effect of the earth's magnetism is found to be several times as great as at sea level.

These results show that a considerable part, at least, of the cosmic rays consists of electrified particles. Some of the cosmic rays, however, are not appreciably affected by the earth's magnetic field. Other types of measurements, such as those of Piccard and Regener in their high-altitude balloon flights and Bothe and Kolhörster's counter experiments, lead us to the conclusion that very little of these rays is in the form of photons, like light, but that there is probably a considerable quantity of radiation in the form of atoms or atomic nuclei of low atomic weight.

A word should be said regarding the tremendous energy represented by individual cosmic rays. Let us take as our unit of energy the electron-volt. About two such units are liberated by burning a hydrogen atom. Two million units appear when radium shoots out an alpha particle. But it requires ten thousand million of these units to make a cosmic ray. Where does this tremendous energy come from? In the answer to this question lies perhaps the solution of the riddle as to how our universe came to be.

Vocabulary

latitude 緯度

due to~ ～のせいで
electrified 電気を帯びた

appreciably 明確に，それとわかるほどに

photon 光子
radiation 放射
atomic nuclei 原子核。単数形は nucleius
atomic weight 原子量
▶ Technical Terms

electron-volt 電子ボルト（エネルギーの単位）
▶ Technical Terms

alpha particle α粒子

Technical Terms

原子量（**atomic weight**） 原子の質量を表す指標で，炭素 12 原子 1 個の原子量を 12 としている。質量数にほぼ等しい。
電子ボルト（**electron-volt**） エネルギーの単位で，1 個の電子が真空中で電位差 1 ボルトの 2 点間で加速されたときに得るエネルギー。また，素粒子物理学などの分野では，素粒子の質量をエネルギーに換算して電子ボルト単位で表すのが一般的だ。

宇宙線の研究は「その現象の微細さ，観測の繊細さ，遠征観測に伴う危険，解析の微妙さ，その壮大な影響において，近代物理学で比類がない」といわれてきた。宇宙線は私たちに何らかの重要なメッセージをもたらしているに違いない。おそらく，自然界の進化や原子核の最深部の構造に関する情報を宇宙線は語っており，私たちはそのメッセージの解読に取り組んでいる。

5年ほど前，ドイツの2人の物理学者ボーテ（Walther Bothe）とコルヘルスター（Werner Kolhörster）は計数管を用いた実験を行い，宇宙線が電気を帯びた粒子であると確信した。この結論が正しいとすると，宇宙線の強度は地球上の場所によって異なるはずだ。地球は巨大な磁石として振る舞い，地球へ飛んでくる荷電粒子は向きを曲げられるからだ。この効果は磁極近くで最小，赤道近くで最大になるので，赤道から両極に向かうにつれて宇宙線の強度は増すはずである。これを検出するために6つの実験が行われたが，決定的なデータは得られなかった。

そこで私たちシカゴ大学のグループはカーネギー協会の資金援助を受けて過去18カ月間に9つの遠征観測を行い，地球上の異なる場所に出向いて，海面レベルからアンデスやヒマラヤの6500m近い高山まで，宇宙線の強さを観測した。カルペとコヴェンという2人の有能な登山家がアラスカのマッキンリー山腹の氷河で命を落としたが，彼らが取得したデータは高緯度地域における最も標高の高い場所での観測値となった。

これらの遠征観測の結果を総合した結果，両極近くの宇宙線強度は赤道近辺に比べて約15%強いことがわかった。さらに，宇宙線の強度は飛来する荷電粒子に地磁気が及ぼす効果を反映して，予想通り緯度によって変わった。また，高地では地磁気のこの効果が海面レベルの数倍に達することがわかった。

これらの結果は，宇宙線の少なくとも相当部分が荷電粒子でできていることを示している。だが，一部の宇宙線は地球磁場から明確な影響を受けない。ピカール（Auguste Piccard）とレゲナー（Erich Regener）の高高度気球による観測やボーテとコルヘルスターの計数管実験など別タイプの計測から，私たちは宇宙線のうち光（光子）の形態を取っている部分はごく一部であり，原子量の小さな原子または原子核がかなりを占めるとの結論に達した。

個々の宇宙線が持つ強大なエネルギーについて一言述べておくべきだろう。エネルギーの単位として電子ボルトを使うことにする。水素が燃えるとき，水素原子1個が解放するエネルギーが約2電子ボルトだ。ラジウムがα粒子1個を放出するときには200万電子ボルトのエネルギーが生じる。これに対し，1個の宇宙線を生み出すには100億電子ボルトが必要になる。こんな膨大なエネルギーがどこからやってくるのだろうか？　この疑問の答えのなかに，私たちの宇宙がどのように生まれたのかという謎に対する答えもおそらく潜んでいる。

The Structure of the Nucleus

原子核の構造（1951年掲載）

M. G. メイヤー（1963年受賞）

マリア・ゲッパート・メイヤー(Maria Göppert Mayer，1906～1972年）はドイツ生まれの米国の物理学者。1963年，「原子核の殻構造に関する発見」によって，ドイツの物理学者ヨハネス・ハンス・ダニエル・イェンゼン（Johannes Hans Daniel Jensen，1907～1973年）とともにノーベル物理学賞を受賞した。ちなみに同年の物理学賞の残り半分は「原子核および素粒子に関する理論への貢献，特に対称性の基本原理の発見とその応用」によってハンガリー出身の米国の物理学者ユージン・ウィグナー（Eugene Wigner，1902～1995年）に授与されている。

メイヤーは1930年に米国人物理学者と結婚して米国に移住，コロンビア大学やシカゴ大学などで研究生活を送った。彼女がイェンゼンとともに提唱した原子核の「シェルモデル」は核構造に関する代表的なモデルで，以下の記事で述べられているように原子核の崩壊や「魔法数」などをうまく説明できる。原子核が原子全体と同様の殻構造をしているという考え方は一般人にも一種のロマンを感じさせる。記事が執筆されたのはノーベル賞を受賞する12年前だ。

核構造に関してはこのほかにも複数のモデルが提唱されている。すべてを説明できる決定版はなく，現在も理論と実験の両面で研究が続いている。抜粋記事の最後の部分に，メイヤーの科学者としての誠実さと謙虚さがにじんでいる。

初出：SCIENTIFIC AMERICAN March 1951
抜粋掲載：SCIENTIFIC AMERICAN July 2012，日経サイエンス2012年11月号

For the atom as a whole modern physicists have developed a useful model based on our planetary system: it consists of a central nucleus, corresponding to the sun, and satellite electrons revolving around it, like planets, in certain orbits. This model, although it leaves many questions still unanswered, has been helpful in accounting for much of the observed behavior of the electrons. The nucleus itself, however, is very poorly understood. Even the question of how the particles of the nucleus are held together has not received a satisfactory answer.

Recently several physicists, including the author, have independently suggested a very simple model for the nucleus. It pictures the nucleus as having a shell structure like that of the atom as a whole, with the nuclear protons and neutrons grouped in certain orbits, or shells, like those in which the satellite electrons are bound to the atom. This model is capable of explaining a surprisingly large number of the known facts about the composition of nuclei and the behavior of their particles.

It is possible to discern some rather remarkable patterns in the properties of particular combinations of protons and neutrons, and it is these patterns that suggest our shell model for the nucleus. One of these remarkable coincidences is the fact that the nuclear particles, like electrons, favor certain "magic numbers."

Every nucleus (except hydrogen, which consists of but one proton) is characterized by two numbers: the number of protons and the number of neutrons. The sum of the two is the atomic weight of the nucleus. The number of protons determines the nature of the atom; thus a nucleus with two protons is always helium, one with three protons is lithium, and so on. A given number of protons may, however, be combined with varying numbers of neutrons, forming several isotopes of the same element.

Vocabulary

our planetary system 私たちの惑星系,つまり太陽系
nucleus 原子核
shell structure 殻構造
proton 陽子
neutron 中性子
composition 組成,構成
discern 識別する,見て取る
coincidence (偶然の)一致
atomic weight 原子量(ここでは質量数)
isotope 同位体
element 元素

Now it is a very interesting fact that protons and neutrons favor even-numbered combinations; in other words, both protons and neutrons, like electrons, show a strong tendency to pair. In the entire list of some 1,000 isotopes of the known elements, there are no more than six stable nuclei made up of an odd number of protons and an odd number of neutrons.

Moreover, certain even-numbered aggregations of protons or neutrons are particularly stable. One of these magic numbers is 2. The helium nucleus, with 2 protons and 2 neutrons, is one of the most stable nuclei known. The next magic number is 8, representing oxygen, whose common isotope has 8 protons and 8 neutrons and is remarkably stable. The next magic number is 20, that of calcium.

The list of magic numbers is: 2, 8, 20, 28, 50, 82 and 126. Nuclei with these numbers of protons or neutrons have unusual stability. It is tempting to assume that these magic numbers represent closed shells in the nucleus, like the electronic shells in the outer part of the atom.

The shell model can explain other features of nuclear behavior, including the phenomenon known as isomerism, which is the existence of long-lived excited states in nuclei. Perhaps the most important application of the model is in the study of beta-decay, i.e., emission of an electron by a nucleus. The lifetime of a nucleus that is capable of emitting an electron depends on the change of spin it must undergo to release the electron. Present theories of beta-decay are not in a very satisfactory state,

Vocabulary

even-numbered 偶数の。even number は偶数

odd number 奇数

aggregation 集合体

tempting 心をそそる

isomerism 核異性
▶ Technical Terms

excited state 励起状態

beta-decay ベータ崩壊
▶ Technical Terms

spin スピン

undergo 経験する

Technical Terms

核異性(**isomerism**) 原子核が衝突などでエネルギーを得て励起状態になり，それがしばらく維持されるものを異性核といい，この現象を核異性と呼んでいる。

ベータ崩壊(**beta-decay**) 原子核の放射性崩壊の一種で，電子と反ニュートリノが放出される。ベータ線は電子のこと。これに伴い，原子核中の中性子 1 個が陽子に変化する。

and it is not easy to check on these theories because only in a few cases are the states of radioactive nuclei known. The shell model can help in this situation, for it is capable of predicting spins in cases in which they have not been measured. Certainly the simple model described here falls short of giving a complete and exact description of the structure of the nucleus. Nonetheless, the success of the model in describing so many features of nuclei indicates that it is not a bad approximation of the truth.

Vocabulary

radioactive 放射性の

fall short of~　～に達しない，できない

approximation 近似

近代の物理学者は，太陽系の姿に基づいて原子の有用なモデルを開発してきた。原子は中心に太陽に相当する原子核があり，惑星が特定の軌道をめぐるように，電子が原子核の周囲を回っている。このモデルは，未解決の多くの謎を残しているとはいえ，周回電子の振る舞いの多くを説明するのに役立ってきた。これに対し，原子核についてはほとんどわかっていない。原子核を構成する粒子がどのように結びつけられているのかという疑問にさえ，十分な答えは得られていない。

最近，著者を含む幾人かの物理学者がそれぞれ独立に，原子核の非常に明快なモデルを提唱した。原子核は原子全体がそうであるように殻構造を持っており，原子核を構成する陽子と中性子は，原子中の電子のように，ある軌道，つまりシェルにグループ化されているという描像だ。このモデルは原子核の組成と核子の振る舞いについて知られている多くの事実を驚くほどよく説明できる。

陽子と中性子の特定の組み合わせが備えている特性には，かなり顕著なパターンがいくつか見て取れる。そして，これらのパターンがそもそも，原子核のシェルモデルを示唆するもとになった。これら注目すべき原子との類似点のひとつは，核子が電子と同様，ある「魔法数」を好むという事実だ。

1個の陽子だけでできている水素の原子核を除き，すべての原子核は2つの数によって特徴づけられる。陽子の数と中性子の数だ。この2つの和が原子核

の原子量（質量数）だ。陽子の数は原子の本質を決め，2個の陽子を持つ原子核は常にヘリウム，3個の陽子を含む原子核はリチウムというように決まる。しかし，陽子の数が同じ原子核でも中性子の数が様々に異なる場合があり，これらは同じ元素の同位体となる。さてここで，陽子と中性子は偶数個になる傾向が強いという興味深い事実がある。言い換えると，陽子も中性子も，電子と同じくペアをなす傾向が強いのだ。既知の元素のすべての同位体およそ1000種のうち，陽子数と中性子数がともに奇数の安定な原子核は6つしかない。

さらに，陽子数や中性子数が偶数である原子核には特に安定なものがある。そうした魔法数の一例は2で，2個の陽子と2個の中性子からなるヘリウムの原子核は最も安定な原子核のひとつだ。次の魔法数は8で，これは酸素に対応し，最も一般的な同位体は8個の陽子と8個の中性子からなり，やはり非常に安定している。次の魔法数は20で，カルシウムに対応する。

すべての魔法数を並べると，2，8，20，28，50，82，126となる。陽子や中性子の数がこれらの魔法数となっている原子核は並外れて安定だ。最外殻の電子が埋まった原子が安定しているように，これらの魔法数は原子核のシェルが埋まった状態に相当するのだと考えたくなる。

原子核のシェルモデルは，原子核中の長寿命の励起状態の存在を示す「核異性」という現象など，原子核のその他の振る舞いも説明できる。シェルモデルの最も重要な応用はおそらく，ベータ崩壊，つまり原子核からの電子の放出に関する研究だろう。電子を放出できる原子核の寿命は，電子の放出に伴って必然的に生じるスピンの変化に依存する。ベータ崩壊に関する現在の理論は必ずしも満足できるものではないし，放射性原子核の励起状態として知られているのはわずかな例だけなので，これらの理論を検証するのも容易ではない。シェルモデルはこの点でも役に立つ。スピンが計測されていない場合でも原子核のスピンを予言できるからだ。ここで述べた簡単なモデルが原子核の構造に関する完全で正確な記述を与えるには不足であるのは明らかだ。それでも，原子核のこれだけ多くの特徴をうまく説明できたということは，このモデルが真実の悪くない近似であることを示している。

The Antiproton

反陽子（1956年掲載）

E. セグレ（1959年受賞）／C. E. ウィーガンド

エミリオ・セグレ（Emilio Gino Segrè，1905～1989年）はイタリア生まれの米国の物理学者。1959年，「反陽子の発見」によって米国の物理学者オーウェン・チェンバレン（Owen Chamberlain，1920～2006年）とともにノーベル物理学賞を受賞した。

セグレは1946年にカルフォルニア大学バークレー校の教授となり，加速器「ベバトロン」を用いた実験で1955年にチェンバレンらとともに反陽子を発見した。以下の記事の共著者ウィーガンド（Clyde E. Wiegand，1915～1996年）は米国の物理学者で，やはり研究チームの一員だ。

この記事は反陽子の発見から1年足らずのうちに執筆されており，研究チームの興奮が伝わってくる。文句なしの大ニュースであり，発見から4年でノーベル賞に輝いた。20世紀半ばの素粒子物理学実験の黄金期を象徴する成果といえるだろう。ベバトロンは当時としては世界最大のシンクロトロンで，この建設を主導したのはローレンス（82ページ参照）だった。

初出：SCIENTIFIC AMERICAN June 1956
抜粋掲載：SCIENTIFIC AMERICAN July 2012，日経サイエンス2012年11月号

A quarter of a century ago P.A.M. Dirac of the University of Cambridge developed an equation, based on the most general principles of relativity and quantum mechanics, which described in a quantitative way various properties of the electron. He had to put in only the charge and mass of the electron—and then its spin, its associated magnetic moment and its behavior in the hydrogen atom followed with mathematical necessity. Its discoverer found, however, that the equation required the existence of both positive and negative electrons: that is, it described not only the known negative electron but also an exactly symmetrical particle which was identical with the electron in every way except that its charge was positive instead of negative.

A few years after Dirac's prediction, Carl D. Anderson of the California Institute of Technology found positive electrons (positrons) among the particles produced by cosmic rays in a cloud chamber. This discovery set physicists off on a new and more formidable search for another hypothetical particle—a search which was finally rewarded only a few months ago.

Dirac's general equation, slightly modified, should be applicable to the proton as well as to the electron. In this instance too it predicts the existence of an antiparticle—an antiproton identical to the proton but with a negative instead of a positive charge.

The question then arose as to how much energy would be needed to create antiprotons in the laboratory

Vocabulary

relativity 相対性
quantum mechanics 量子力学
charge 電荷
spin スピン
magnetic moment 磁気モーメント
symmetrical 対称的な
positron 陽電子
cosmic ray 宇宙線
cloud chamber 霧箱
▶ Technical Terms
formidable 手強い
applicable 適用できる
proton 陽子
antiparticle 反粒子
▶ Technical Terms
antiproton 反陽子

Technical Terms

霧箱(**cloud chamber**)　気体中に霧を発生させて荷電粒子の飛跡を観測する装置。1897年に英国の物理学者チャールズ・ウィルソン(Charles Thomson Rees Wilson)が発明，1927年のノーベル物理学賞を受賞した。霧箱は20世紀半ばまで粒子の観測や宇宙線の検出に多用されたが，現在では用いられず，博物館的な存在だ。

反粒子(**antiparticle**)　ある素粒子について，電荷の正負だけが逆の粒子のこと。マイナスの電荷を持った反陽子，電子と逆にプラスの電気を持つ陽電子など。

with an accelerator. Because an antiproton can be created only in a pair with a proton, we need at least the energy equivalent to the mass of two protons (i.e., about two billion electron volts). However, we need much more than two Bev in the proposed laboratory experiment. To convert energy into particles we must concentrate the energy at a point; this is best accomplished by hurling a high-energy particle at a target—e.g., a proton against a proton. After the collision we shall have four particles: the two original protons plus the newly created proton-antiproton pair. Each of the four will emerge from the collision with a kinetic energy amounting to about one Bev. Thus the generation of an antiproton takes two Bev (creation of the proton-antiproton pair) plus four Bev (the kinetic energy of the four emerging particles). It was with these numbers in mind that the Bevatron at the University of California was designed.

When the Bevatron began to bombard a target made of copper with six-Bev protons, the next problem was to detect and identify any antiprotons created. A plan for the search was devised by Owen Chamberlain, Thomas Ypsilantis and the authors of this article. The plan was based on three properties which could conveniently be determined. First, the stability of the particle meant that it should live long enough to pass through a long apparatus. Second, its negative charge could be identified by the direction of deflection of the particle by an applied magnetic field, and the magnitude of its charge could be gauged by the amount of ionization it produced along its path. Third, its mass could be calculated from the curve of its trajectory in a given magnetic field if its velocity was known.

Vocabulary

accelerator 加速器
equivalent to~　～と同等の
electron volt　電子ボルト（エネルギーの単位）
Bev　billion electron volts（10億電子ボルト）の略。
concentrate 集中する
hurl 投げつける，発射する
kinetic energy 運動エネルギー
Bevatron ベバトロン
▶ Technical Terms
detect 検出する
identify 同定する
conveniently うまい具合に
deflection 屈折，運動方向の変化
ionization イオン化，電離

Technical Terms

ベバトロン（**Bevatron**）　1954年に米国のカリフォルニア大学に建造された陽子シンクロトロン。加速エネルギー60億電子ボルト（6GeV，米国流の言い方では6BeV）を誇る当時としては世界最大の加速器で，BeV級の加速器という意味でこの名がついた。

When the discovery of the antiproton was announced last October, 60 of them had been recorded, at an average rate of about four to each hour of operation of the Bevatron. They had passed all the tests which we had preordained before the start of the experiment. We were quite gratified by the comment of a highly esteemed colleague who had just finished an important and difficult experiment on mesons. After examining our tests, he said, "I wish that my own experiments on mu mesons were as convincing as this." At this time several long-standing bets on the existence of the antiproton started to be paid. The largest we know of was for $500. (We were not personally involved.)

Vocabulary

preordain あらかじめ定める

meson 中間子
▶ Technical Terms

mu meson　ミュー中間子
▶ Technical Terms

Technical Terms

中間子(**meson**)　1個のクォークと1個の反クォークが結びついてできている粒子で，素粒子間の「強い相互作用」を媒介する。湯川秀樹が1934年に理論的に導入し(1949年ノーベル物理学賞)，後にパイ中間子として発見された。ほかにもいくつかの種類がある。

ミュー中間子(**mu meson**)　いまでいうミュー粒子(ミューオン，**muon**)のこと。ミュー粒子が最初に発見された1937年，質量が近いことから中間子と考えられてこう名づけられたが，実際には中間子の仲間ではなく，電子などとともに「レプトン」と総称される粒子グループに属する。

四半世紀前，英ケンブリッジ大学のディラック（P.A.M. Dirac）は相対論と量子力学の最も一般的な原理に基づき，電子の様々な特性を量的に記述する方程式を作り出した。この方程式に電子の電荷と質量を入れるだけで，電子のスピンとそれに付随する磁気モーメント，水素原子のエネルギー準位が数学的な必然性から導かれる。だがディラックは，その方程式がプラスの電子とマイナスの電子の両方を必要とすることに気づいた。つまり，この方程式は負電荷を帯びた既知の電子を記述するだけでなく，その電荷がマイナスではなくプラスであることを除けば電子とまったく同じである対称的な粒子も記述する。

ディラックの予言から数年後，カリフォルニア工科大学のアンダーソン（Carl D. Anderson）が，宇宙線が霧箱に作り出した粒子のなかにプラス電荷の電子（陽電子）を発見した。この発見が刺激となり，物理学者たちはもう 1 つの仮想上の粒子を探す手強い試みに取り組んだ。この探索は，つい数カ月前にとうとう実を結んだ。

ディラックの方程式を少し変えると，電子だけでなく陽子に適用できるようになる。この場合もやはり，1 つの反粒子の存在が予言される。電荷がプラスではなくマイナスであることを除けば陽子とまったく同じ「反陽子」だ。

次に問題となるのは，加速器実験で反陽子を作り出すのにどれだけのエネルギーが必要かという点だ。反陽子は陽子とペアになった形でのみ作られるので，少なくとも陽子 2 個の質量に相当するエネルギー（約 20 億電子ボルト）が必要だ。だが，実際の加速器実験では 2Bev よりずっと大きなエネルギーが必要になる（訳注：Bev は billion electron volts の略で，現在でいう GeV のこと）。エネルギーを粒子に転換するには，エネルギーを一点に集中させなければならない。それには高エネルギーの粒子を標的にぶつけるのが最善で，例えば陽子を陽子と衝突させる。衝突後には 4 個の粒子が得られるはずだ。もともとの陽子 2 個と，新たに生じた陽子・反陽子のペアである。これら 4 個の粒子はそれぞれ約 1Bev の運動エネルギーを持って衝突点から飛び出してくるだろう。だから，1 個の反陽子を生成するには 2Bev（陽子・反陽子ペアの生成）プラス 4Bev（飛び出てくる 4 個の粒子の運動エネルギー）が必要だ。カリフォルニア大学にあるベバトロン（Bevatron）はこの数字を念頭に設計された。

ベバトロンで銅の標的に6Bevの陽子をぶつける実験が始まり，次の問題は生成された反陽子を検出・同定することに移った。チェンバレン（Owen Chamberlain）とイプシランティス（Thomas Ypsilantis），そして私たちこの記事の著者が1つの計画を考案した。この計画は実験でうまく測定できる3つの特性に基づいている。第1に，反陽子の安定性は，反陽子が長い装置を通り抜けるに十分な寿命を持っているかどうかで評価できる。第2に，反陽子の負電荷は粒子が印加磁場のなかで曲げられる方向によって識別でき，その電荷の大きさは粒子の軌跡に沿って生じたイオン化の量によって測ることができる。第3に，反陽子の質量は，その速度がわかっている場合，一定の磁場中での軌跡の曲がり具合から計算できる。

反陽子の発見が発表された昨年10月時点で，60個が記録されていた。ベバトロンの運転1時間あたりに平均で約4個が生じたことになる。これらの60例は実験開始前に定めていたすべての検証試験にパスした。私たちは，中間子に関する重要で困難な実験を終えたばかりのある高名な科学者のコメントを聞いて，非常にうれしく思った。彼は私たちの検証を吟味した後，「ミュー中間子に関する私自身の実験もこれと同じ説得力があったらいいのにと思う」と述べたという。現在，反陽子の存在に関する長年の賭のいくつかについて，賞金の払い戻しが始まっている。私たち著者が知るなかで最高額は500ドルだ（私たち自身は賭に参加していない）。

Optical Masers

光メーザー（1961 年掲載）

A. L. ショーロー（1981 年受賞）

アーサー・レナード・ショーロー（Arthur Leonard Schawlow, 1921 ～ 1999 年）は米国の物理学者。1981 年,「レーザー分光学への貢献」によって米国の物理学者ニコラス・ブルームバーゲン（Nicolaas Bloembergen, 1920 年～）とノーベル物理学賞を受賞した。ちなみに同年の物理学賞の半分は「高分解能光電子分光法の開発」によってスウェーデンの物理学者カイ・シーグバーン（Kai Manne Börje Siegbahn, 1918 ～ 2007 年）に授与された。

以下の記事は, 1960 年に米国の物理学者チャールズ・ハード・タウンズ（Charles Hard Townes, 1915 ～ 2015 年）がレーザーの発振に成功したことを受けて，その翌年に執筆されたものだ。「光メーザー」という記事の表題は時代を感じさせる。メーザーはいわばマイクロ波のレーザーで，レーザーより先に，やはりタウンズらが発振に成功していた(1954 年)。その“光版”が登場したばかりの 1961 年, レーザーという呼び名はまだ定着していなかったのだ（この言葉そのものは 1959 年に初めて使われてはいた)。

ちなみにタウンズは 1964 年,「量子エレクトロニクス分野の基礎研究およびメーザー・レーザー原理に基づく振動子・増幅器の構築」によって，ロシアの物理学者ニコライ・バソフ（Nikolay Gennadiyevich Basov, 1922 ～ 2001 年）およびアレクサンドル・ミハイロヴィチ・プロホロフ（Aleksandr Mikhailovich Prokhorov, 1916 ～ 2002 年)とともにノーベル物理学賞を受賞した。一方, ショーローら 1981 年の受賞者の功績は高度な分光解析を発展させたことにある。

初出：SCIENTIFIC AMERICAN June 1961
抜粋掲載：SCIENTIFIC AMERICAN July 2012, 日経サイエンス 2012 年 11 月号

For at least half a century communications engineers have dreamed of having a device that would generate light waves as efficiently and precisely as radio waves can be generated. The contrast in purity between the electromagnetic waves emitted by an ordinary incandescent lamp and those emitted by a radio-wave generator could scarcely be greater. Radio waves from an electromagnetic oscillator are confined to a fairly narrow region of the electromagnetic spectrum and are so free from "noise" that they can be used for carrying signals. In contrast, all conventional light sources are essentially noise generators that are unsuited for anything more than the crudest signaling purposes. It is only within the last year, with the advent of the optical maser, that it has been possible to attain precise control of the generation of light waves.

Although optical masers are still very new, they have already provided enormously intense and sharply directed beams of light. These beams are much more monochromatic than those from other light sources.

The optical maser is such a radically new kind of light source that it taxes the imagination to canvass its possible applications. Message-carrying, of course, is the most obvious use and the one that is receiving the most technological attention. Signaling with light, although it has been used by men since ancient times, has been limited by the weakness and noisiness of available light sources. An ordinary light beam can be compared to a pure, smooth carrier wave that has already been modulated with noise by short bursts of light randomly emitted by the individual

Vocabulary

radio wave ラジオ波，電波
electromagnetic wave 電磁波
incandescent lamp 白熱電球
oscillator 発振器
electromagnetic spectrum 電磁スペクトラム
crude 粗雑な
optical maser 光メーザー
▶ Technical Terms
monochromatic 単色性の
tax 課する，強いて求める
canvass 詳細に調べる，描く
carrier wave 搬送波
modulate 変調する

Technical Terms

光メーザー（**optical maser**） いまでいうレーザー（laser）のこと。メーザーは波長と位相のそろったマイクロ波で，Microwave Amplification by Stimulated Emission of Radiation（誘導放出によるマイクロ波増幅）の略称。レーザーは波長がより短い光の帯域にこれを発展させたもので，Light Amplification by Stimulated Emission of Radiation（誘導放出による光増幅）の略だ。名前の通り，どちらも誘導放出という現象に基づいている。

atoms in the light source. The maser, on the other hand, can provide an almost ideally smooth wave, carrying nothing but what one puts on it.

If suitable methods of modulation can be found, coherent light waves should be able to carry an enormous volume of information. This is so because the frequency of light is so high that even a very narrow band of the visible spectrum includes an enormous number of cycles per second; the amount of information that can be transmitted is directly proportional to the number of cycles per second and therefore to the width of the band. In television transmission the carrier wave carries a signal that produces an effective bandwidth of four megacycles. A single maser beam might reasonably carry a signal with a frequency, or bandwidth, of 100,000 megacycles, assuming a way could be found to generate such a signal. A signal of this frequency could carry as much information as all the radiocommunication channels now in existence. It must be admitted that no light beam will penetrate fog, rain or snow very well. Therefore to be useful in earthbound communication systems light beams will have to be enclosed in pipes.

Vocabulary

modulation 変調
coherent コヒーレントな
▶ Technical Terms
band 帯域
proportional to~ 〜に比例して
bandwidth 帯域幅
reasonably 正当に，無理なく
radiocommunication 無線通信
earthbound 現実的な

Technical Terms

コヒーレントな(**coherent**)　「波長と位相がそろった」という意味。位相(波の山谷のパターン)のそろった波形が空間的・時間的に長く保たれており，干渉性(コヒーレンス)に優れている。レーザーが代表例。

少なくともこの半世紀，通信技術者は電波と同様の効率と正確さで光波を生み出す装置を夢見てきた。通常の白熱電球が放出する電磁波と発振器が生み出す電磁波とでは，その純度が恐ろしいほど違う。発振器からの電波は電磁スペクトラムのかなり狭い領域に閉じ込められており，“ノイズ”がほとんどないので信号の伝送に使える。これと対照的に，従来の光源は基本的にノイズ発生器であり，ごく粗雑な信号伝達以上のものには向いていない。光波の発生を正確にコントロールできるようになったのはつい昨年，光メーザーが出現してからだ。

光メーザーはまだ生まれたての新技術だが，すでに非常に強くて指向性の鋭い光ビームが生み出されている。これらの光ビームは他のどの光源よりも単色性に優れている。

非常に革新的な光源なので，どんな応用が可能か，想像をいやがおうでも刺激する。もちろんメッセージ伝送は最も明らかな用途であり，寄せられている技術的関心も高い。光による信号伝達は，人類が古代から使ってきた方法ではあるが，利用できる光源の弱さとノイズの多さに制限されてきた。通常の光線は，光源の個々の原子がランダムに発する短い光のバーストによるノイズのせいで，純粋で滑らかな搬送波がすでに変調されてしまったものだといえる。メーザーは対照的に，ほぼ理想的な滑らかな光波を発生でき，そこに乗せた信号だけを伝えられる。

もし適切な変調法が見つかれば，コヒーレントな光波は莫大な量の情報を伝送できるはずだ。それが可能なのは，光の周波数がとても高く，可視光スペクトルのごく狭い帯域でも，1秒あたりに膨大なサイクル数を含んでいることによる。伝送可能な情報の量はサイクル数に比例し，したがって帯域の幅に比例する。テレビ放送の場合，搬送波は4メガサイクルの帯域幅に相当する信号を伝送している。メーザーのビームなら1本で，周波数つまり帯域幅が10万メガサイクルの信号を送れるかもしれない（そんな信号を生成する方法が見つかったとしての話だが）。この周波数の信号は，現存の無線通信チャンネルすべてが伝えているのと同量の情報を伝えられるだろう。ただし，霧や雨，雪に邪魔されない光ビームはないことも認めねばならない。だから，実用的な通信システムとして利用するには，光ビームをパイプのなかに封入する必要があるだろう。

X-ray Stars

X 線星（1967 年掲載）

R. ジャコーニ（2002 年受賞）

リカルド・ジャコーニ（Ricardo Giacconi，1931 年～）はイタリア出身の米国の宇宙物理学者。2002 年，「宇宙 X 線源の発見を導いた天体物理学への先駆的貢献」によってノーベル物理学賞を受賞した。同年の物理学賞の半分は「天体物理学への先駆的貢献，特に宇宙ニュートリノの検出」で日本の物理学者・小柴昌俊（1926 年～）および米国の化学者・物理学者レイモンド・デービス（Raymond Davis，1914 ～ 2006 年）に授与された。

宇宙には太陽を含め X 線を発する天体がたくさん存在するが，地球の大気が X 線を吸収するため地上では観測できない。観測機器を大気圏外に打ち上げて X 線をとらえる試みが 20 世紀半ばに始まり，ジャコーニはその先駆者として知られる。

以下の記事はノーベル賞を受賞する 35 年も前に執筆されたもので，1960 年代の X 線天文学の黎明期を伝えている。宇宙を見る新たな手段を得た興奮が感じられるほか，現在までに日米欧が多数の X 線観測衛星を展開して先進的な観測を進めるなど，半世紀でこの分野が大きく進展したことを改めて実感できる。

一方，2002 年に同時受賞した小柴昌俊氏は超新星爆発に伴うニュートリノを地下に設置された巨大検出装置「カミオカンデ」でとらえ，「ニュートリノ天文学」という新領域を拓いた（141 ページ「ニュートリノの質量の発見」も参照）。高エネルギーのニュートリノは宇宙線の起源など大きな謎の解明につながると期待されており，今後はニュートリノ天文学が花開くだろう。

初出：SCIENTIFIC AMERICAN December 1967
抜粋掲載：SCIENTIFIC AMERICAN July 2012, 日経サイエンス 2012 年 11 月号

Although interstellar space is suffused with radiation over the entire electromagnetic spectrum, from the extremely short waves of gamma rays and X rays to the very long radio waves, relatively little of the cosmic radiation reaches the earth's surface. Our atmosphere screens out most of the wavelengths. In particular the atmosphere is completely opaque to wavelengths shorter than 2,000 angstrom units. Hence X radiation from space can be detected only by sending instruments to the outer regions of our atmosphere in balloons or rockets.

As rocket flights and opportunities to send up instrumented payloads became more frequent, Bruno B. Rossi of the Massachusetts Institute of Technology suggested an X-ray survey of the sky, and a group of us at American Science and Engineering, Inc., undertook the study.

The instrumented Aerobee rocket was launched at the White Sands Missile Range at midnight on June 18, 1962. Our experiment had been prepared by Herbert Gursky, F. R. Paolini and me, with Rossi's collaboration. Some time before the rocket arrived at its peak altitude 225 kilometers (140 miles) above the earth's surface, doors opened to expose the detectors. With the rocket spinning on its axis, the detectors scanned a 120-degree belt of the sky, including the position of the moon.

The telemeter signals from the detectors showed no indication of any X radiation coming from the moon. From the direction of the constellation Scorpio in the southern sky, however, the detectors revealed the presence

Vocabulary

interstellar space 星間空間
be suffused with~ ～で満たされる
radiation 放射
spectrum スペクトラム
▶ Technical Terms
opaque 不透明
angstrom オングストローム
▶ Technical Terms

altitude 高度

telemeter 遠隔計測器

Scorpio 星座のさそり座。constellationは星座

Technical Terms

スペクトラム(**spectrum**)　本来は光の周波数(波長)別の強さを表したグラフのことだが，ここでは「周波数帯域の広がり」といった意味で使っている。スペクトル。

オングストローム(**angstrom**)　長さの単位で，10^{-10}m つまり0.1nmのこと。水素原子の大きさが約1オングストロームで，可視光の波長は数千オングストロームになることから，分光学などの分野ではかつて広く使われていた。現在ではナノメートル単位を用いるのが原則だ。

of an intense source of X rays. The intensity registered by the counters was a million times greater than one would expect (on the basis of the sun's rate of X-ray emission) to arrive from any distant cosmic source!

Three months of close study of the records verified that the radiation was indeed X rays (two to eight angstroms in wavelength), that it came from outside the solar system and that the source was roughly in the direction of the center of our galaxy. What kind of object could be emitting such a powerful flux of X rays?

We made two additional rocket surveys at different times of the year (in October, 1962, and June, 1963) that narrowed down the location of the strong X-ray source by triangulation, and we found that it was not actually in the galactic center. Meanwhile Herbert Friedman and his collaborators at the Naval Research Laboratory succeeded in locating the position of the source within a two-degree arc in the sky, which suggested that the X-ray emitter was a single star rather than a large collection of them.

By this time the evidence that the source was a discrete object had become so strong that we named it Sco (for Scorpius) X-1. One might have expected that an object pouring out so much energy in X radiation would be distinctly visible as at least a rather bright star. The region of the source was barren, however, of conspicuous stars.

The problem then was to identify the X-ray star among the visible stars at the indicated location. The position of Sco X-1 was known only within about one degree, and in its region of the sky there are about 100 13th-magnitude stars in each square degree. A detailed analysis of the new data was made to pinpoint the position more closely. This analysis narrowed the location to two equally probable positions where the star might be found.

Vocabulary

solar system 太陽系
object 天体
narrow down 絞り込む
triangulation 三角測量
discrete 個別の, 分離した
Scorpius さそり座。Scorpioに同じ
pour out 流し出す, 放出する
distinctly はっきりと
be barren of ~ ~に乏しい
conspicuous 目立つ
magnitude 恒星の明るさの等級

Given these positions, the Tokyo Astronomical Observatory and the Mount Wilson and Palomar Observatories made a telescopic search for Sco X-1. The Tokyo astronomers found the X-ray star immediately, and within a week the Palomar observers confirmed the identification.

Now that Sco X-1 can be examined with optical telescopes, it is beginning to yield some striking new information. The most provocative fact is that this star emits 1,000 times more energy in X rays than in visible light, a situation astronomers had never anticipated from their studies of the many varieties of known stars. There are indications that the X-ray emission of Sco X-1 is equal to the total energy output of the sun at all wavelengths.

Vocabulary

identification 確認, 同定

optical telescope 光学望遠鏡

provocative 刺激的な

星間空間は極めて短波長のガンマ線やX線から長波長の電波まで，電磁気的な全スペクトラムにわたる放射で満たされているものの，これら宇宙放射のうち地表に達するのは比較的わずかだ。大気がほとんどの波長を遮蔽してしまう。特に波長が2000オングストローム（1オングストロームは0.1nm）よりも短い放射はまったく透過しない。だから，宇宙からのX線放射を検出するには，計測器を気球やロケットによって大気の外層部に持って行かねばならない。

ロケットの飛行が増え，計測機器を搭載して打ち上げる機会も増えたことを受け，マサチューセッツ工科大学のロッシ（Bruno B. Rossi）はX線での掃天観測を提案し，私たちアメリカン・サイエンス・アンド・エンジニアリング社のグループがそれを引き受けた。

観測機器を搭載したエアロビー・ロケットは1962年6月18日の深夜にホワイトサンズ・ミサイル実験場から打ち上げられた。観測機器はロッシの協力のもと，私とガースキー（Herbert Gursky），パオリーニ（F. R. Paolini）が開発した。ロケットが地上約225kmの最高高度に達するまでのある時点で，扉が開いて検出器が露出した。ロケットはその軸を中心に回転しているので，検出器は天空を帯状に120°にわたって走査し，その範囲には月の位置も含まれる。

検出器からの信号は月からX線が来ている兆しはないことを示していた。だが，南天のさそり座の方向に強いX線源の存在が明らかになった。計器が記録したその強度は，いかなる遠方のX線源に予想される強度（太陽のX線放射に基づいて推定された値）よりも，何と100万倍も大きかった！

データを3カ月かけて詳しく調べた結果，この放射が確かにX線であり（波長は2～8オングストローム），太陽系の外から来たもので，発生源は天の川銀河のほぼ中心の方向にあることが確かめられた。こんな強力なX線を放出しているのは，いったいどんな天体なのか？

私たちはその後，異なる季節を選んで2回の追加観測（1962年10月と1963年6月）を行い，この強いX線源の位置を三角測量によって絞り込んで，実際には銀河中心にあるのではないことを突き止めた。一方，米海軍研究所のフリードマン（Herbert Friedman）らはこの X 線源の天空上の位置を 2° 角以内の精度で決定することに成功した。これは X 線源が星の大集団ではなく，単一の星であることを示している。

この段階で，X 線源が独立の天体であることが確かになったので，私たちはこれを「さそり座 X-1」と名づけた。これほど大量の X 線放射エネルギーを放出している天体なら，肉眼でもかなり明るい星として見えると期待してもおかしくはない。しかし，X 線源の領域は明るい星に乏しかった。

そうなると，次の問題はその領域に見えている星のなかからこの X 線星を特定することだ。さそり座 X-1 の位置はざっと 1° 角の領域としてわかっているだけで，この領域には約 100 個の 13 等星がある。位置をもっと正確に特定するため，新しいデータを詳細に解析した。これによって，X 線星が見つかりそうな位置が 2 つに絞り込まれた。

これをもとに，東京天文台とウィルソン山天文台，パロマー天文台がさそり座 X-1 を望遠鏡で探索した。東京の天文学者がすぐにこの X 線星を発見し，1 週間以内にパロマー天文台がこれを確認した。

これでさそり座 X-1 を光学望遠鏡で調べられるようになり，いくつかの驚くべき情報が得られ始めている。最も刺激的な事実は，この星が可視光の 1000 倍以上のエネルギーを X 線の形で放出していることで，これは様々な星を研究してきた天文学者もまるで予想していなかった状況だ。さそり座 X-1 の X 線放射は太陽が全波長で放出している総エネルギーに匹敵すると考えられる。

How a Supernova Explodes

超新星爆発のメカニズム（1985年掲載）

H. A. ベーテ（1967年受賞）／ G. ブラウン

ハンス・アルブレヒト・ベーテ（Hans Albrecht Bethe，1906～2005年）は米国の物理学者。1967年，「原子核反応理論への貢献，特に星の内部におけるエネルギー生成に関する発見」によってノーベル物理学賞を単独で受賞した。太陽をはじめ恒星が核融合反応によってエネルギーを生み出していることを1939年に示した業績が最も有名で，ほかにも原子核反応理論に関する数々の重要な成果を残している。

以下の記事は，大質量の恒星が一生の最後に起こす超新星爆発について述べた内容だ。ノーベル賞受賞から20年近くたったこの分野の重鎮が，超新星爆発のメカニズムを平易に解説している。1980年代の知見に基づいているが，30年後の現在でもそのまま通用する解説といえる。

なお，共著者のブラウン（Gerald Brown，1926年～）は米国の物理学者で，執筆当時はニューヨーク州立大学ストーニーブルック校の教授。

初出：SCIENTIFIC AMERICAN May 1985, サイエンス 1985年7月号
抜粋掲載：SCIENTIFIC AMERICAN July 2012, 日経サイエンス 2012年11月号

A supernova begins as a collapse, or implosion; how does it come about, then, that a major part of the star's mass is expelled? At some point the inward movement of stellar material must be stopped and reversed; an implosion must be transformed into an explosion.

Through a combination of computer simulation and theoretical analysis a coherent view of the supernova mechanism is beginning to emerge. It appears the crucial event in the turnaround is the formation of a shock wave that travels outward.

When the center of the core reaches nuclear density, it is brought to rest with a jolt. This gives rise to sound waves that propagate back through the medium of the core, rather like the vibrations in the handle of a hammer when it strikes an anvil. The waves slow as they move out through the homologous core, both because the local speed of sound declines and because they are moving upstream against a flow that gets steadily faster. At the sonic point they stop entirely. Meanwhile additional material is falling onto the hard sphere of nuclear matter in the center, generating more waves. For a fraction of a millisecond the waves collect at the sonic point, building up pressure there. The bump in pressure slows the material falling through the sonic point, creating a discontinuity in velocity. Such a discontinuous change in velocity constitutes a shock wave.

At the surface of the hard sphere in the heart of the star infalling material stops suddenly but not instantaneously. Momentum carries the collapse beyond the point of equilibrium, compressing the central core to a density even higher than that of an atomic nucleus. We call this point the instant of "maximum scrunch." After the

Vocabulary

supernova 超新星
▶ Technical Terms
collapse 重力崩壊
▶ Technical Terms
implosion 爆縮
come about 起こる，生じる
explosion 爆発

coherent 首尾一貫した

turnaround 転換
shock wave 衝撃波

nuclear 原子核の
bring to rest 停止させる
give rise to~ ～を生じる

anvil 鉄床（かなとこ）

homologous 同質な，均質な

sonic point 音速点

bump 出っ張り，高まり
discontinuity 不連続

momentum 運動量
equilibrium 平衡

scrunch 詰め込む

Technical Terms

超新星（**supernova**） 質量の大きな恒星が一生の最後に起こす大爆発のこと。
重力崩壊（**collapse**） 星の重力が星内部の圧力を上回って，星が無限に収縮すること。超新星爆発の前段階にあたる。

maximum scrunch the sphere of nuclear matter bounces back, like a rubber ball that has been compressed. The bounce sets off still more sound waves, which join the growing shock wave.

Shock wave differs from a sound wave in two respects. First, a sound wave causes no permanent change in its medium; when the wave has passed, the material is restored to its former state. The passage of a shock wave can induce large changes in density, pressure and entropy. Second, a sound wave—by definition—moves at the speed of sound. A shock wave moves faster, at a speed determined by the energy of the wave. Hence once the pressure discontinuity at the sonic point has built up into a shock wave, it is no longer pinned in place by the infalling matter. The wave can continue outward, into the overlying strata of the star. According to computer simulations, it does so with great speed, between 30,000 and 50,000 kilometers per second.

After the outer layers of a star have been blown off, the fate of the core remains to be decided. The explosion of lighter stars presumably leaves behind a stable neutron star. In Wilson's calculations any star of more than about 20 solar masses leaves a compact remnant of more than two solar masses. It would appear that the remnant will become a black hole, a region of space where matter has been crushed to infinite density.

Vocabulary

set off 引き起こす，放つ

medium 媒質
restore 回復する

entropy エントロピー

overlying overlie（～の上に重なる）の現在分詞

neutron star 中性子星
▶ Technical Terms

remnant 残骸
black hole ブラックホール
infinite 無限大の

Technical Terms

中性子星（**neutron star**） 超新星爆発の結果，その中心部にできる超高密度の天体。中性子が主な成分で，半径は 10km 程度と小さいが太陽と同程度の質量が詰め込まれていると考えられている。

超新星は重力崩壊，つまり爆縮として始まる。では，どのようにして星の質量の大半が飛び散るのだろうか。内側に向かっていた物質の運動がどこかで止まり，そして逆転しなければならない。爆縮は爆発へと転換されなければならない。

コンピューターシミュレーションと理論解析を組み合わせることで，超新星のメカニズムの統一的な理解が得られようとしている。方向転換を起こすのに決定的な役割を果たしているのは，星の表面へと伝わっていく衝撃波の形成であることがわかってきた。

コアの中心が原子核と同じ密度に達したとき，中心部は急激な衝撃とともに停止させられる。これによって，中心から外側へと逆向きに伝わる音波が生じる。鉄床をたたいたときに金づちの柄に伝わってくる振動のようなものだ。この波は，同質なコアの中を外側へ伝わるにつれて遅くなっていく。その理由は2つあり，1つは各場所での音速が減少するため，もう1つは，しだいにスピードを増す物質の流れに逆らってさかのぼっていくためだ。音速点で音波は完全に止まってしまう。一方，中心の核物質の硬い球の上にはさらに物質が落下し，新たな音波を作る。1ミリ秒もたたない間に波が音速点に集まり，そこで圧力を生み出す。この圧力が，音速点を通過してくる物質の落下を遅らせ，速度の不連続を引き起こす。こうした速度の不連続な変化が衝撃波を作り出す。

落下してきた物質は星の中心にある硬い球の表面で急に止まるが，瞬時に止まるわけではない。運動量は平衡点を超えて重力崩壊を進ませ，中心のコアは原子核の密度よりもさらに高い密度まで圧縮される。私たちはこの点を“最大圧搾”の瞬間と呼んでいる。最大圧搾の後，核物質の球は押しつぶされていたゴムボールが元に戻るように跳ね返る。この跳ね返りによって，さらに多くの音波が生まれ，音速点で成長しつつある衝撃波に加わる。

衝撃波は2つの点で音波とは異なる。第1に，音波が原因で媒質が永久的に変化することはない。音波が通り過ぎると，物質は以前の状態に戻る。これに対し，衝撃波が通過すると密度，圧力，エントロピーに大きな変化が引き起こされる。第2に，音波は当然ながら，音速で伝わる。衝撃波はもっと速く伝わり，その速度は波のエネルギーによって決まる。したがって，音速点での圧力の不連続がひとたび衝撃波に成長すると，もはや物質の落下によって衝撃波をその場にとどめておくことはできない。波は外へ広がり続け，星の上層へ伝わっていく。コンピューターシミュレーションによれば，秒速3万～5万kmという猛スピードで広がる。

星の外層が吹き飛ばされた後のコアはどうなるのだろうか。比較的軽い星の爆発は，おそらく安定な中性子星を後に残すと考えられる。ウィルソン（James R. Wilson）の計算によると，太陽質量の20倍以上の星であれば，太陽質量の2倍以上のコンパクトな残骸を残す。この残骸はブラックホールになると思われる。ブラックホールとは，物質が無限大の密度まで押しつぶされた空間領域のことである。

The Higgs Boson

ヒッグス・ボソンは実在するか（1986年掲載）

M. J. G. フェルトマン（1999年受賞）

マルティヌス・フェルトマン（Martinus J. G. Veltman，1931年～）はオランダの物理学者。1999年，「電弱相互作用の量子構造の解明」によって，やはりオランダの理論物理学者であるゲラルド・トホーフト（Gerardus 't Hooft，1946年～）とノーベル物理学賞を受賞した。電弱理論（電磁気力と素粒子の「弱い相互作用」と呼ばれる力を統一的に記述する理論）をはじめ，素粒子物理学の研究をリードしてきた理論家だ。

ノーベル賞受賞の13年前に執筆された以下の記事は，素粒子物理学の基本的な理論体系である「標準モデル」において存在が予想されながら実験的確認が最後まで取れなかったヒッグス粒子（ヒッグス・ボソン）に関するものだ。その後2012年になって，欧州合同原子核研究機構（CERN）の大型ハドロン衝突型加速器LHCによってようやく確認されたが，記事掲載当時の1980年代はまだ手探りに近い状態だったことが読み取れる。

抜粋記事中で著者がヒッグス粒子検出の手立てとして期待している米国の超電導超大型衝突型加速器SSCはその後，建設が中止されて実現しなかった。LHCが実験を開始したのは2008年だ。理論家と実験物理学者たちが長い年月をかけて総力で取り組んだ努力を思わずにはいられない。

なお，ヒッグス粒子の発見を受け，これを理論的に提唱した英国の物理学者ピーター・ヒッグス（Peter Higgs, 1929年～）とベルギーの物理学者フランソワ・アングレール（François Englert，1932年～）に2013年のノーベル物理学賞が贈られた。

初出：SCIENTIFIC AMERICAN November 1986, サイエンス 1987年1月号
抜粋掲載：SCIENTIFIC AMERICAN July 2012, 日経サイエンス 2012年11月号

The Higgs boson, which is named after Peter W. Higgs of the University of Edinburgh, is the chief missing ingredient in what is now called the standard model of elementary processes: the prevailing theory that describes the basic constituents of matter and the fundamental forces by which they interact. According to the standard model, all matter is made up of quarks and leptons, which interact with one another through four forces: gravity, electromagnetism, the weak force and the strong force. The strong force, for instance, binds quarks together to make protons and neutrons, and the residual strong force binds protons and neutrons together into nuclei. The electromagnetic force binds nuclei and electrons, which are one kind of lepton, into atoms, and the residual electromagnetic force binds atoms into molecules. The weak force is responsible for certain kinds of nuclear decay. The influence of the weak force and the strong force extends only over a short range, no larger than the radius of an atomic nucleus; gravity and electromagnetism have an unlimited range and are therefore the most familiar of the forces.

In spite of all that is known about the standard model, there are reasons to think it is incomplete. That is where the Higgs boson comes in. Specifically, it is held that the Higgs boson gives mathematical consistency to the standard model, making it applicable to energy ranges beyond the capabilities of the current generation of particle accelerators but that may soon be reached by future accelerators. Moreover, the Higgs boson is thought to generate the masses of all the fundamental particles; in a manner of speaking, particles "eat" the Higgs boson to gain weight.

Vocabulary

Higgs boson ヒッグス粒子，ヒッグス・ボソン
standard model 標準モデル
quark クォーク ▶ Technical Terms
lepton レプトン ▶ Technical Terms
weak force 弱い相互作用
strong force 強い相互作用
proton 陽子
neutron 中性子
nuclei nucleus（原子核）の複数形
nuclear decay 原子核の崩壊
extend 影響が及ぶ，届く
consistency 一貫性，矛盾がないこと
particle accelerator 加速器
fundamental particle 基本粒子

Technical Terms

クォーク（**quark**） 陽子や中性子，中間子などを構成している基本粒子。強い相互作用の担い手で，全部で６種類が知られている。
レプトン（**lepton**） 強い相互作用をしない粒子の一群で，電子やミュー粒子，ニュートリノなど。

The biggest drawback of the Higgs boson is that so far no evidence of its existence has been found. Instead a fair amount of indirect evidence already suggests that the elusive particle does not exist. Indeed, modern theoretical physics is constantly filling the vacuum with so many contraptions such as the Higgs boson that it is amazing a person can even see the stars on a clear night! Although future accelerators may well find direct evidence of the Higgs boson and show that the motivations for postulating its existence are correct, I believe things will not be so simple. I must point out that this does not mean the entire standard model is wrong. Rather, the standard model is probably only an approximation—albeit a good one—of reality.

Forces among elementary particles are investigated in high-energy-physics laboratories by means of scattering experiments. A beam of electrons might, for instance, be scattered off a proton. By analyzing the scattering pattern of the incident particles, knowledge of the forces can be gleaned.

The electroweak theory successfully predicts the scattering pattern when electrons interact with protons. It also successfully predicts the interactions of electrons with photons, with W bosons [particles that make the weak field felt] and with particles called neutrinos. The theory runs into trouble, however, when it tries to predict the interaction of W bosons with one another. In particular, the theory indicates that at sufficiently high energies the probability of scattering one W boson off another W boson is greater than 1. Such a result is clearly

Vocabulary

drawback 欠点

elusive とらえにくい

contraption 新案, 珍奇な仕掛け

motivation 動機
postulate 仮定する

approximation 近似

elementary particle 素粒子

scattering experiments 衝突実験。scattering は散乱
incident 入射の

electroweak theory 電弱理論 ▶ Technical Terms
interact 相互作用する

W boson W ボソン, W 粒子

neutrino ニュートリノ

Technical Terms

電弱理論(**electroweak theory**)　電磁気力と素粒子の「弱い相互作用」を統一的に記述する理論のこと。1960 年代から 70 年代に確立した。136 ページのワインバーグ(Steven Weinberg)に関する説明も参照。

nonsense. The statement is analogous to saying that even if a dart thrower is aiming in the opposite direction from a target, he or she will still score a bull's-eye.

It is here that the Higgs boson enters as a savior. The Higgs boson couples with the W bosons in such a way that the probability of scattering falls within allowable bounds: a certain fixed value between 0 and 1. In other words, incorporating the Higgs boson in the electroweak theory "subtracts off" the bad behavior.

Armed with the insight that the Higgs boson is necessary to make the electroweak theory renormalizable, it is easy to see how the search for the elusive particle should proceed: [W bosons] must be scattered off one another at extremely high energies, at or above one trillion electron volts (TeV). The necessary energies could be achieved at the proposed 20-TeV Superconducting Super Collider (SSC), which is currently under consideration in the U.S. If the pattern of the scattered particles follows the predictions of the renormalized electroweak theory, then there must be a compensating force, for which the Higgs boson would be the obvious candidate. If the pattern does not follow the prediction, then the [W bosons] would most likely be interacting through a strong force, and an entire new area of physics would be opened up.

Vocabulary

score a bull's-eye 金的を射る，大当たりを取る

savior 救世主

bounds 範囲

renormalizable くりこみ可能
▶ Technical Terms

electron volt 電子ボルト（エネルギーの単位）

Superconducting Super Collider 超電導超大型衝突型加速器

Technical Terms

くりこみ可能（**renormalizable**） 「くりこみ」は場の量子論で計算結果が無限大に発散して意味をなさなくなるのを防ぐ数学的な手法のことで，この処理を適用できることを「くりこみ可能」という。くりこみの理論を確立した朝永振一郎とシュウィンガー（Julian Seymour Schwinger），ファインマン（Richard Phillips Feynman）は1965年のノーベル物理学賞を受賞。

英エディンバラ大学のヒッグス（Peter W. Higgs）の名にちなんでヒッグス・ボソンと呼ばれる粒子は，素過程の標準モデルと呼ばれる理論の主要な“欠けた鎖”だ。この理論は物質の基本的な構成要素とその間に働く基本的な力を記述する。標準モデルによれば，すべての物質はクォークとレプトンからできており，4つの力，重力・電磁力・弱い力・強い力を通じて相互作用する。例えば強い力はまずクォークを束縛して陽子や中性子を作り，余った力で陽子や中性子を束縛して原子核を作る。電磁力は，原子核と，レプトンの一種である電子を結びつけて原子を作り，余りの力で原子どうしを結びつけて分子を作る。弱い力はある種の原子核の崩壊の原因になる。弱い力と強い力の影響が及ぶ距離はごく短く，原子核の半径を超えない。一方，重力と電磁力は無限大の距離にまで効き，4つの力の中でもなじみ深い力となっている。

標準モデルについてはよくわかっているにもかかわらず，これが不完全と考えられる理由がいくつかある。ここで登場するのがヒッグス・ボソンだ。ヒッグス・ボソンは標準モデルの数学的なつじつまを合わせ，現世代の加速器の到達範囲は超えるが近い将来の加速器で達しうるようなエネルギー範囲にまで，この理論を適用できるようにする，と考えられている。さらに，ヒッグス・ボソンはすべての基本粒子の質量を作り出すと考えられている。いわば，粒子はヒッグス・ボソンを“食べて”質量を獲得するのだ。

ヒッグス・ボソンの最大の欠点は，それの存在証拠が今まで何ひとつ見いだされていない点である。むしろ，少なからぬ間接的証拠が，この捕捉しにくい粒子が存在しないことを示唆している。実際，現代の理論物理学はヒッグス・ボソンのような珍妙な粒子をたくさん真空に詰め込んできたわけで，それでもまだ晴れた夜空に星が見えるのが不思議なくらいだ！　将来の加速器がヒッグス・ボソンを直接に検証し，その存在を仮定した動機が正しいことが判明するかもしれないが，事はそう単純ではないと私は思う。標準モデルがすべて間違いだという意味ではない。標準モデルはおそらく，よくできてはいるが，現実のひとつの近似にすぎないのだろう。

粒子間の力は高エネルギーの衝突実験によって調べられている。例えば電子のビームが陽子によって散乱される実験だ。入射粒子の散乱のパターンを解析することで，力に関する知見を探り出すことができる。

電弱理論は電子が陽子と相互作用する際の散乱パターンを正しく予言する。電子と光子，電子とWボソン（弱い力を媒介する粒子），電子とニュートリノの相互作用も正しく記述する。しかし，Wボソンどうしの相互作用を予言しようとすると立ち往生する。十分に高いエネルギーでWボソンどうしの散乱を計算すると，その確率が1を超えてしまうのだ。これは明らかに無意味で，ダーツを標的と反対の向きに投げても的中するというようなものだ。

ここでヒッグス・ボソンが救世主として現れる。この粒子はWボソンと相互作用して，散乱の確率が許容範囲内に落ちるように，つまり0と1の間の値になるようにする。言い換えると，電弱理論の中で好ましからぬ振る舞いを“引き去る”のである。

電弱力をくりこみ可能にするにはヒッグス・ボソンが必要であるという洞察を使えば，このとらえにくい粒子をどう探索すればよいか，容易に見極めがつく。Wボソンどうしは極めて高いエネルギー〔1TeV（1兆電子ボルト）かそれ以上〕で散乱を起こすはずだ。必要なエネルギー領域は，米国で現在検討中である20TeVの超電導超大型衝突型加速器SSCで達成できるだろう。もし散乱粒子の様子がくりこんだ電弱理論の予言にしたがっていたら，何かの力によってそうなっているはずであり，ヒッグス・ボソンはその明らかな候補だ。散乱の様子が予言と合わないならば，Wボソンもおそらく強い力を通じて相互作用していることになり，物理学のまったく新しい分野が開けてくるだろう。

Accurate Measurement of Time

究極の時間測定技術（1993年掲載）

W. M. イタノ／ N. F. ラムゼー（1989年受賞）

ノーマン・フォスター・ラムゼー（Norman Foster Ramsey，1915～2011年）は米国の物理学者。1989年，「分離振動場法の開発およびその水素メーザーや原子時計への応用」によってノーベル物理学賞を受賞した。同年の物理学賞の残り半分は「イオントラップ法の開発」によって米国の物理学者ハンス・デーメルト（Hans G. Dehmelt，1922年～）およびドイツの物理学者ヴォルフガング・パウル（Wolfgang Paul，1913～1993年）に授与された。

ラムゼーの授賞理由にある分離振動場法は「ラムゼー共鳴」として知られる現象に基づくもので，高精度の原子時計を実現するのに欠かせない。ノーベル賞受賞から4年後に書かれた以下の記事では，正確な時間測定を目指す最先端の技術を解説している。イオンや電気的に中性な原子を真空容器のなかで中空に静止させ（ノーベル賞を同時受賞したイオントラップ法はそのひとつ），そこにレーザー光を当てるという技術的に高度な操作が必要となる。かなり専門的な技術だが，正確な時間標準を追求してやまない科学者・技術者の情熱を感じさせる。

共著者のイタノ（Wayne M. Itano, 1951年～）は米国立標準技術研究所（NIST）の物理学者。

初出：SCIENTIFIC AMERICAN July 1993, 日経サイエンス 1993年9月号
抜粋掲載：SCIENTIFIC AMERICAN July 2012, 日経サイエンス 2012年11月号

New technologies, relying on the trapping and cooling of atoms and ions, offer every reason to believe that clocks can be 1,000 times more precise than existing ones.

One of the most promising depends on the resonance frequency of trapped, electrically charged ions. Trapped ions can be suspended in a vacuum so that they are almost perfectly isolated from disturbing influences. Hence, they do not suffer collisions with other particles or with the walls of the chamber.

Two different types of traps are used. In a Penning trap, a combination of static, nonuniform electric fields and a static, uniform magnetic field holds the ions. In a radio frequency trap (often called a Paul trap), an oscillating, nonuniform electric field does the job. Workers at Hewlett-Packard, the Jet Propulsion Laboratory in Pasadena, Calif., and elsewhere have fabricated experimental standard devices using Paul traps. The particles trapped were mercury 199 ions. The maximum Qs [a measure of relative energy absorption and loss] of trapped-ion standards exceed 10^{12}. This value is 10,000 times greater than that for current cesium beam clocks [the higher the Q, the more stable the clock].

During the past few years, there have been spectacular developments in trapping and cooling neutral atoms, which had been more difficult to achieve than trapping ions. Particularly effective laser cooling results from the use of three pairs of oppositely directed laser-cooling beams along three mutually perpendicular paths. A moving atom is then slowed down in whatever direction it moves. This effect gives rise to the designation "optical molasses." Neutral-atom traps can store higher densities of

Vocabulary

trap 捕捉する

resonance frequency 共鳴周波数

isolate 隔離する

Penning trap ペニング・トラップ
electric field 電場
magnetic field 磁場
radio frequency trap RFトラップ，ポール・トラップ

fabricate 作製する

mercury 水銀

cesium セシウム

neutral 電気的に中性な

laser cooling レーザー冷却
▶ Technical Terms

perpendicular 直交する

molasses 糖蜜，シロップ

Technical Terms

レーザー冷却(**laser cooling**) 気体分子にレーザー光を当てることで，その温度を極低温にまで冷却する方法のこと。分子の上下・左右・前後からレーザーを当てると，分子の運動方向と逆向きの力が働き，分子が静止に向かう。

atoms than can ion traps, because ions, being electrically charged, are kept apart by their mutual repulsion. Other things being equal, a larger number of atoms result in a higher signal-to-noise ratio.

The main hurdle in using neutral atoms as frequency standards is that the resonances of atoms in a trap are strongly affected by the laser fields. A device called the atomic fountain surmounts the difficulty. The traps capture and cool a sample of atoms that are then given a lift upward so that they move into a region free of laser light. The atoms then fall back down under the influence of gravity. On the way up and again on the way down, the atoms pass through an oscillatory field. In this way, resonance transitions are induced, just as they are in the separated oscillatory field beam apparatus.

Much current research is directed toward laser-cooled ions in traps that resonate in the optical realm, where frequencies are many thousands of gigahertz. Such standards provide a promising basis for accurate clocks because of their high Q. Investigators at NIST have observed a Q of 10^{13} in the ultraviolet resonance of a single laser-cooled, trapped ion. This value is the highest Q that has ever been seen in an optical or microwave atomic resonance.

The anticipated improvements in standards will increase the effectiveness of the current uses and open the way for new functions. Only time will tell what these uses will be.

Vocabulary

repulsion 反発
result in~ ～に帰着する，～を生じる
signal-to-noise ratio 信号対雑音比，SN 比

frequency standards 周波数標準
resonance 共鳴，共振
atomic fountain 原子泉
surmount 障害を乗り越える，切り抜ける
free of~ ～のない

resonance transition 共鳴遷移

原子やイオンを捕捉して冷却する新技術が登場したおかげで，現在より1000倍も精密な時計が実現可能と考えられる。

最も有望なのは，トラップ（捕捉）した荷電イオンの共鳴周波数を利用する方法だ。トラップしたイオンを真空中に浮遊させ，妨害するものからほぼ完全に隔離することができる。つまり，個々のイオンは他の粒子や容器の壁との衝突を受けない。

2種類のトラップが使われている。「ペニング・トラップ」は不均一な静電場と均一な静磁場によってイオンを閉じ込める。「RFトラップ」（ポール・トラップとも呼ばれる）は不均一な高周波電場を使う。ヒューレット・パッカードやジェット推進研究所などで，ポール・トラップを使った試験的な標準装置が作られている。捕捉された粒子は水銀199のイオンだ。トラップイオン標準のQ値（粒子がどれほどよくエネルギーを保つかを示す指標）は最大で10^{12}を超える。この値は現在のセシウムビーム時計の1万倍だ（Q値が大きいほど安定な時計となる）。

ここ数年で，イオンよりもずっと難しい「中性原子のトラップと冷却」が劇的に進歩した。特に互いに直交する3つの軸に反対向きのレーザービームをそれぞれ配置することで，効果的なレーザー冷却が実現した。こうすると，どの方向に動いている原子でも減速される。この状態は“光の糖蜜状態”と呼ばれる。中性原子トラップはイオントラップに比べ，高い密度で原子を蓄えられる。イオンは帯電しているので，相互に反発しあって離れているからだ。他の条件が同じなら，原子の個数が多いほどSN比（信号対雑音比）が高くなる。

中性原子を周波数標準に使う上での最大の障害は，トラップ中の原子の共鳴がレーザーの場から強く影響を受けることだ。この問題の解決には「原子泉」と呼ばれる方法が使われる。捕捉・冷却された原子を上の方へ打ち上げて，レーザー光のない空間に移す。その後，原子は重力のために落下する。原子は上がるときと下がるときに振動場を通過する。こうして，別のビームによって個別の振動場を作る装置と同様に，共鳴遷移が誘導される。

現在の多くの研究は，可視光の領域，つまり周波数が数千 GHz 以上の領域で共鳴するトラップ中でイオンをレーザー冷却することに向けられている。こうした標準は Q 値が高いため，正確な時計の有望な基礎になる。NIST（米国立標準技術研究所）の研究者たちは，レーザー冷却した 1 個の捕捉イオンの紫外光共鳴で 10^{13} という高い Q 値を観測した。この値は光やマイクロ波の原子共鳴でこれまでに観測された最高の値だ。

周波数標準は今後も改良が進み，現在の応用例の効率がさらに高まるとともに，新たな機能の実現にも道が開けるだろう。それがどんなものになるかは，時のみが語ってくれるだろう。

Life in the Universe

宇宙の中の生命（1994年掲載）

S. ワインバーグ（1979年受賞）

スティーブン・ワインバーグ（Steven Weinberg, 1933年～）は米国の物理学者。「素粒子間に働く弱い相互作用と電磁相互作用を統一した理論への貢献，特に弱中性カレントの予想」によって，パキスタンの物理学者アブドゥス・サラム（Abdus Salam, 1926～1996年）および米国の物理学者シェルドン・グラショー（Sheldon Lee Glashow，1932年～）とともに1979年のノーベル物理学賞を受賞した。いわゆる電弱理論を完成させた功績で，「ワインバーグ＝サラム理論」あるいは「グラショー＝ワインバーグ＝サラム理論」とも呼ばれている。

以下の記事は電弱理論そのものではなく，「この宇宙が物理的に極めて特殊な条件を満たすように“微調整”されているからこそ生命が存在しうる」という，ある意味で不可解な謎に関するものだ。抜粋記事中に登場する「真空のエネルギー」はいわゆる暗黒エネルギー（ダークエネルギー）のこと。宇宙の加速膨張をもたらしている原因とされるが，正体不明なうえ，理論値と観測から導かれる値とが甚だしく食い違っており，現代物理学・宇宙論の最大の謎となっている。

ノーベル賞受賞から15年後，押しも押されもせぬ知の巨人が，さらにスケールの大きな謎について考察したエッセイといえる。

なお，日経サイエンス1994年12月号に掲載された初出記事の翻訳は当時京都大学教授だった益川敏英氏（2008年ノーベル物理学賞受賞）らによる。

初出：SCIENTIFIC AMERICAN October 1994, 日経サイエンス 1994年12月号
抜粋掲載：SCIENTIFIC AMERICAN July 2012, 日経サイエンス 2012年11月号

Life as we know it would be impossible if any one of several physical quantities had slightly different values. The best known of these quantities is the energy of one of the excited states of the carbon 12 nucleus. There is an essential step in the chain of nuclear reactions that build up heavy elements in stars. In this step, two helium nuclei join together to form the unstable nucleus of beryllium 8, which sometimes before fissioning absorbs another helium nucleus, forming carbon 12 in this excited state. The carbon 12 nucleus then emits a photon and decays into the stable state of lowest energy. In subsequent nuclear reactions carbon is built up into oxygen and nitrogen and the other heavy elements necessary for life. But the capture of helium by beryllium 8 is a resonant process, whose reaction rate is a sharply peaked function of the energies of the nuclei involved. If the energy of the excited state of carbon 12 were just a little higher, the rate of its formation would be much less, so that almost all the beryllium 8 nuclei would fission into helium nuclei before carbon could be formed. The universe would then consist almost entirely of hydrogen and helium, without the ingredients for life.

Opinions differ as to the degree to which the constants of nature must be fine-tuned to make life necessary. There are independent reasons to expect an excited state of carbon 12 near the resonant energy. But one constant does seem to require an incredible fine-tuning: it is the vacuum energy, or cosmological constant, mentioned in connection with inflationary cosmologies.

Vocabulary

excited state 励起状態
▶ Technical Terms
nucleus 原子核
element 元素

fission 核分裂する

resonant process 共鳴過程

vacuum energy 真空のエネルギー
▶ Technical Terms
cosmological constant 宇宙定数
▶ Technical Terms
inflationary cosmology インフレーション宇宙論

Technical Terms

励起状態(**excited state**)　原子核や原子などの量子系は，異なるエネルギー状態を取ることができる。最も安定なのはエネルギーが最低の「基底状態」だが，外部からエネルギーを得て，より高いエネルギー状態になったのが励起状態。これらのエネルギー状態は「準位」と呼ばれ，離散的な値を取る。
真空のエネルギー（**vacuum energy**）　真空そのものが持つとされるエネルギーで，宇宙膨張を加速させているいわゆる暗黒エネルギー（ダークエネルギー）のこと。
宇宙定数(**cosmological constant**)　もともとはアインシュタインが定常的な宇宙を実現するために一般相対性理論の重力方程式に追加した「宇宙項」のこと。この定数の意味合いは暗黒エネルギーと同じであることが後にわかった。

Although we cannot calculate this quantity, we can calculate some contributions to it (such as the energy of quantum fluctuations in the gravitational field that have wavelengths no shorter than about 10^{-33} centimeter). These contributions come out about 120 orders of magnitude larger than the maximum value allowed by our observations of the present rate of cosmic expansion. If the various contributions to the vacuum energy did not nearly cancel, then, depending on the value of the total vacuum energy, the universe either would go through a complete cycle of expansion and contraction before life could arise or would expand so rapidly that no galaxies or stars could form.

Thus, the existence of life of any kind seems to require a cancellation between different contributions to the vacuum energy, accurate to about 120 decimal places. It is possible that this cancellation will be explained in terms of some future theory. So far, in string theory as well as in quantum field theory, the vacuum energy involves arbitrary constants, which must be carefully adjusted to make the total vacuum energy small enough for life to be possible.

All these problems can be solved without supposing that life or consciousness plays any special role in the fundamental laws of nature or initial conditions. It may be that what we now call the constants of nature actually vary from one part of the universe to another. (Here "different parts of the universe" could be understood in various senses. The phrase could, for example, refer to different local expansions arising from episodes of inflation in which the fields pervading the universe took different values or else to the different quantum-mechanical "worldtracks" that arise in some versions of quantum cosmology.) If this were the case, then it would not be surprising to find that life is possible in some parts of the universe, though perhaps not in most.

Vocabulary

quantum fluctuation 量子ゆらぎ
gravitational field 重力場
cosmic expansion 宇宙膨張
~decimal places 小数点以下～桁
string theory ひも理論
quantum field theory 場の量子論
arbitrary 任意の
initial conditions 初期条件
inflation （宇宙の）インフレーション

Naturally, any living beings who evolve to the point where they can measure the constants of nature will always find that these constants have values that allow life to exist. The constants have other values in other parts of the universe, but there is no one there to measure them. Still, this presumption would not indicate any special role for life in the fundamental laws, any more than the fact that the sun has a planet on which life is possible indicates that life played a role in the origin of the solar system.

いくつかの物理定数は，そのうちどれかの値が少しでも違っていたなら，私たちが知る生命は存在できなかった。そうした物理量で最もよく知られたものは，炭素 12 のある励起状態のエネルギー準位である。恒星中で重元素を合成する一連の原子核反応には，欠くことのできない段階がある。それは 2 つのヘリウム原子核が結合してベリリウム 8 の不安定な原子核を作る段階で，このベリリウム 8 は時に分裂するよりも先にヘリウム原子核をもう 1 つ吸収し，前述の励起状態にある炭素 12 になる。この炭素 12 の原子核はその後，光子を 1 個放出して最低エネルギーの安定状態へと崩壊する。その後に続く原子核反応の中で，炭素は酸素や窒素など，生命に必要な重元素に変化していく。しかし，ベリリウム 8 によるヘリウムの捕獲は共鳴過程であり，その反応確率はそれにかかわる原子核のエネルギー準位のところで鋭いピークを持った関数である。もし炭素 12 の励起状態のエネルギー準位がほんの少し高かったら，この過程の確率はずっと小さくなり，ほぼすべてのベリリウム 8 原子核はヘリウム原子核に分裂してしまって炭素にはならないだろう。すると宇宙はほぼすべてが水素とヘリウムからなり，生命の原料は存在しないことになる。

生命を必然的なものとするために自然定数がどの程度まで精密に調整されていなければならないかについては，意見が分かれている。共鳴エネルギーの近くに炭素 12 の励起状態があることは，いくつかの別の理由から期待される。しかし，ある自然定数は信じられないような精密な調整を要求されているらしい。それは真空のエネルギーである。宇宙定数とも呼ばれるこの量は，インフレーション宇宙論に関連して語られる。

この量を完全に計算することはできないが，それに寄与するものの一部は計算されている（重力場中の 10^{-33}cm より長い波長を持つ量子論的ゆらぎのエネルギーなど)。これらの寄与は観測された現在の宇宙膨張率から許される最大値と比べても，ほぼ 120 桁も大きい。真空のエネルギーに対する様々な寄与がほとんど打ち消しあっているのでない限り，真空の全エネルギーの値によって，宇宙は生命が発生する以前に膨張と収縮を繰り返すか，銀河や恒星が形成される暇もないほど素早く膨張してしまうだろう。

このように，生命の存在は真空のエネルギーに対する様々な寄与の合計が小数点以下 120 桁の精度で打ち消し合うことを要求しているようだ。何らかの理論によって将来，この打ち消し合いが説明される可能性はある。今のところ，ひも理論や場の量子論では，真空のエネルギーは任意の定数を伴っており，それらの定数は真空の全エネルギーが生命を可能にするほど小さくなるように注意深く調節される必要がある。

生命や意識が自然の基本法則や宇宙の初期条件に対して何か特別の役割を果たすと考えなくても，これらの問題は解決可能だ。私たちが「自然定数」と呼んでいるものが，実は宇宙の異なった場所では別の値になっているのかもしれない（ここで「宇宙の異なった場所」は様々な意味に理解できる。例えば，宇宙を満たしていた場の値が異なっていたために異なるインフレーションが起こって生じた局所的な膨張部分であるとか，ある種の量子論的宇宙論で考えられている異なった量子力学的「世界の軌跡」などが考えられるだろう)。もしこれが本当なら，宇宙の大部分ではおそらく生命が存在できなくても，一部で生命が可能であるのは驚くに当たらないだろう。

当然ながら，自然定数を測定できるまでに進化した生物なら，それらの定数が生命の存在を許す値になっていることを必ず見いだすだろう。宇宙の別の場所では自然定数が違った値を取っているが，そこにはそれを測定する者はいない。それでも，この推論は基本法則に対して生命が特別の役割を果たしたことを意味しない。太陽が生命が存在可能な惑星を持つからといって，太陽系の起源において生命が何か役割を果たしたとはいえないのと同様だ。

Detecting Massive Neutrinos

ニュートリノの質量の発見（1999 年掲載）

梶田隆章（2015 年受賞）／**戸塚洋二**／ **E. カーンズ**

梶田隆章（1959 年〜）は東京大学宇宙線研究所の所長を務める物理学者・天文学者で，2015 年に「ニュートリノ振動の発見」によってカナダの物理学者アーサー・マクドナルド（Arthur B. McDonald，1943 年〜）とともにノーベル物理学賞を受賞した。

ニュートリノ振動とは，素粒子のニュートリノ（3 タイプある）が飛行中に別のタイプのニュートリノに変化してはまた元に戻る現象で，質量ゼロと想定されてきたこの素粒子がわずかながら質量を持つ証拠となる。梶田氏らは日本のニュートリノ観測施設「スーパーカミオカンデ」を用いてこの現象をとらえた。マクドナルドはカナダにある別の装置によってこれを確認した。

以下の記事（掲載したのは一部）はスーパーカミオカンデの研究・観測チームがニュートリノ振動の確認を発表した翌年，1999 年に執筆された。当時，梶田氏は宇宙線研究所の助教授。共著者の戸塚洋二氏（1942 〜 2008 年）は当時の宇宙線研究所の所長で，このプロジェクトを主導し，長らくノーベル賞候補とされたが残念ながら死去した。カーンズ（Edward Kearns）はボストン大学の教授で，梶田氏とともに観測データ解析の共同リーダーを務めた。

初出：SCIENTIFIC AMERICAN August 1999, 日経サイエンス 1999 年 10 月号

One man's trash is another man's treasure. For a physicist, the trash is "background"—some unwanted reaction, probably from a mundane and well-understood process. The treasure is "signal"—a reaction that we hope reveals new knowledge about the way the universe works. Case in point: over the past two decades, several groups have been hunting for the radioactive decay of the proton, an exceedingly rare signal (if it occurs at all) buried in a background of reactions caused by elusive particles called neutrinos. The proton, one of the main constituents of atoms, seems to be immortal. Its decay would be a strong indication of processes described by Grand Unified Theories that many believe lie beyond the extremely successful Standard Model of particle physics. Huge proton-decay detectors were placed deep underground, in mines or tunnels around the world, to escape the constant rain of particles called cosmic rays. But no matter how deep they went, these devices were still exposed to penetrating neutrinos produced by the cosmic rays.

The first generation of proton-decay detectors, operating from 1980 to 1995, saw no signal, no signs of proton decay—but along the way the researchers found that the supposedly mundane neutrino background was not so easy to understand. One such experiment, Kamiokande, was located in Kamioka, Japan, a mining town about 250 kilometers (155 miles) from Tokyo (as the neutrino flies). The name stood for "Kamioka Nucleon Decay Experiment." Scientists there used sensitive detectors to peer into ultrapure water, waiting for the telltale flash of a proton decaying.

Vocabulary

background 背景，バックグラウンド
mundane ありふれた
radioactive decay 放射性崩壊
proton 陽子
elusive つかまえにくい
neutrino ニュートリノ
constituent 構成要素
immortal 不変の，不死の
Grand Unified Theory 大統一理論 ▶ Technical Terms
Standard Model 標準モデル
detector 検出器
cosmic ray 宇宙線
penetrate 透過する
Kamiokande カミオカンデ
peer 透かして熟視する

Technical Terms

大統一理論（**Grand Unified Theory**） 自然界の4つの力（強い力，弱い力，電磁気力，重力）のうち，重力を除く3つを統一的に記述する理論を大統一理論（GUT）ということが多い。これら素粒子物理学の標準モデルが扱う3つの力のうち，弱い力と電磁気力を一貫して記述する電弱統一理論が1960年代に成立し，さらに強い力まで統合しようとする挑戦が続いている。

Such an event would have been hidden, like a needle in a small haystack, among about 1,000 similar flashes caused by neutrinos interacting with the water's atomic nuclei. Although no proton decay was seen, the analysis of those 1,000 reactions uncovered a real treasure—tantalizing evidence that the neutrinos were unexpectedly fickle, changing from one species to another in midflight. If true, that phenomenon was just as exciting and theory-bending as proton decay.

Neutrinos are amazing, ghostly particles. Every second, 60 billion of them, mostly from the sun, pass through each square centimeter of your body (and of everything else). But because they seldom interact with other particles, generally all 60 billion go through you without so much as nudging a single atom. A detector as large as Kamiokande catches only a tiny fraction of the neutrinos that pass through it every year.

Neutrinos come in three flavors, corresponding to their three charged partners in the Standard Model: the electron and its heavier relatives, the muon and the tau particle. An electron-neutrino interacting with an atomic nucleus can produce an electron; a muon-neutrino makes a muon; a tau-neutrino, a tau. For most of the seven decades since neutrinos were first posited, physicists have assumed that they are massless. But if they can change from one flavor to another, quantum theory indicates that they most likely have mass. And in that case, these ethereal particles could collectively outweigh all the stars in the universe.

Building a Bigger Neutrino Trap

As is so often the case in particle physics, the way to make progress is to build a bigger machine. Super-Kamiokande, or Super-K for short, took the basic design of Kamiokande and scaled it up by about a factor of 10. An array of light-sensitive detectors looks in toward the center of 50,000 tons of water whose protons may decay or

Vocabulary

haystack 干し草の山
interact 相互作用する
nuclei nucleus（原子核）の複数形
tantalizing 興味をそそる
fickle 変わりやすい
amazing 驚くべき
nudge そっと突く
flavor フレーバー，素粒子における変種
electron 電子
muon ミュー粒子
tau particle タウ粒子
electron-neutrino 電子ニュートリノ
muon-neutrino ミューニュートリノ
tau-neutrino タウニュートリノ
ethereal ごく軽い，触知できない，エーテルのような
Super-Kamiokande スーパーカミオカンデ

get struck by a neutrino. In either case, the reaction creates particles that are spotted by means of a flash of blue light known as Cherenkov light, an optical analogue of a sonic boom, discovered by Pavel A. Cherenkov in 1934. Much as an aircraft flying faster than the speed of sound produces a shock wave of sound, an electrically charged particle (such as an electron or muon) emits Cherenkov light when it exceeds the speed of light in the medium in which it is moving. This motion does not violate Einstein's theory of relativity, for which the crucial velocity is *c*, the speed of light in a vacuum. In water, light propagates 25 percent slower than *c*, but other highly energetic particles can still travel almost as fast as *c* itself. Cherenkov light is emitted in a cone along the flight path of such particles.

In Super-K, the charged particle generally travels just a few meters and the Cherenkov cone projects a ring of light onto the wall of photon detectors. The size, shape and intensity of this ring reveal the properties of the charged particle, which in turn tell us about the neutrino that produced it. We can easily distinguish the Cherenkov patterns of electrons from those of muons: the electrons generate a shower of particles, leading to a fuzzy ring quite unlike the crisper circle from a muon. From the Cherenkov light we also measure the energy and direction of the electron or muon, which are decent approximations to those of the neutrino.

Super-K cannot easily identify the third type of neutrino, the tau-neutrino. Such a neutrino can only interact with a nucleus and make a tau particle if it has enough energy. A muon is about 200 times as heavy as an electron; the tau about 3,500 times. The muon mass is well within the range of atmospheric neutrinos, but only a tiny

Vocabulary

Cherenkov light チェレンコフ光

shock wave 衝撃波
electrically charged particle 荷電粒子
medium 媒質

theory of relativity 相対性理論
propagate 伝播する

project 投射する

intensity 強さ

fuzzy ぼやけた
crisp はっきりした

decent まずまずの
approximation 近似

atmospheric neutrino 大気ニュートリノ
▶ Technical Terms

Technical Terms

大気ニュートリノ（**atmospheric neutrino**）　地表に降り注ぐニュートリノのうち，宇宙線が大気に衝突してできたパイ中間子やミュー粒子が崩壊する過程で生じたニュートリノのこと。

fraction are at tau energies, so most tau-neutrinos in the mix will pass through Super-K undetected.

One of the most basic questions experimenters ask is, "How many?" We have built a beautiful detector to study neutrinos, and the first task is simply to count how many we see. Hand in hand with this measurement is the question, "How many did we expect?" To answer that, we must analyze how the neutrinos are produced.

Super-K monitors atmospheric neutrinos, which are born in the spray of particles when a cosmic ray strikes the top of our atmosphere. The incoming projectiles (called primary cosmic rays) are mostly protons, with a sprinkling of heavier nuclei such as helium or iron. Each collision generates a shower of secondary particles, mostly pions and muons, which decay during their short flight through the air, creating neutrinos. We know roughly how many cosmic rays hit the atmosphere each second and roughly how many pions and muons are made in each collision, so we can predict how many neutrinos to expect.

Tricks with Ratios

Unfortunately, this estimate is only accurate to 25 percent, so we take advantage of a common trick: often the ratio of two quantities can be better determined than either quantity alone. For Super-K, the key is the sequential decay of a pion to a muon and a muon-neutrino, followed by the muon's decay to an electron, an electron-neutrino and another muon-neutrino. No matter how many cosmic rays are falling on the earth's atmosphere, or how many pions they produce, there should be about two muon-neutrinos for every electron-neutrino. The calculation is more complicated than that and involves

Vocabulary

projectile 投射物
primary cosmic rays 一次宇宙線 ▶ Technical Terms
sprinkling 少量
pion パイ中間子
sequential 一連の, 続いて起こる

Technical Terms

一次宇宙線(**primary cosmic rays**)　宇宙空間を飛び交っている宇宙線がそのまま地球に到達したもの。これに対し，一次宇宙線が大気と衝突して生まれた高エネルギー粒子のことは二次宇宙線と呼ぶ。

computer simulations of the cosmic ray showers, but the final predicted ratio is accurate to 5 percent, providing a much better benchmark than the individual numbers of particles do.

After counting neutrinos for almost two years, the Super-K team has found that the ratio of muon-neutrinos to electron-neutrinos is about 1.3 to 1 instead of the expected 2 to 1. Even if we stretch our assumptions about the flux of neutrinos, how they interact with the nuclei and how our detector responds to these events, we cannot explain such a low ratio—unless neutrinos are changing from one type into another.

We can play the ratio trick again to test this surprising conclusion. The clue to our second ratio is to ask how many neutrinos should arrive from each possible direction. Primary cosmic rays fall on the earth's atmosphere almost equally from all directions, with only two effects spoiling the uniformity. First, the earth's magnetic field deflects some cosmic rays, especially the low-energy ones, skewing the pattern of arrival directions. Second, cosmic rays that skim the earth at a tangent make showers that do not descend deep into the atmosphere, and these can develop differently from those that plunge straight in from above.

But geometry saves us: if we "look" up into the sky at some angle from the vertical and then down into the ground at the same angle, we should "see" the same number of neutrinos coming from each direction. Both sets of neutrinos are produced by cosmic rays hitting the atmosphere at the same angle; it is just that in one case the collisions happen overhead and in the other they are partway around the world. To use this fact, we select neutrino events of sufficiently high energy (so their parent cosmic ray was not deflected by the earth's magnetic field) and then divide the number of neutrinos going up by the

Vocabulary

benchmark 比較の基準点

stretch 広げる
assumption 仮定
flux 流量, フラックス

spoil 害する
magnetic field 磁場
deflect 向きを変える
skew ゆがめる
skim かすめて進む
tangent 接線方向
develop 発達する
plunge 飛び込む

geometry 幾何学
vertical 垂直線

partway 途中で

number going down. This ratio should be exactly 1 if no neutrinos are changing flavor.

We saw essentially equal numbers of high-energy electron-neutrinos going up and down, as expected, but only half as many upward muon-neutrinos as downward ones. This finding is the second indication that neutrinos are changing identity. Moreover, it provides a clue to the nature of the metamorphosis. The upward muon-neutrinos cannot be turning into electron-neutrinos, because there is no excess of upward electron-neutrinos. That leaves the tau-neutrino. The muon-neutrinos that become tau-neutrinos pass through Super-K without interaction, without detection.

Vocabulary

metamorphosis 変身, 変性

ある人にとってのゴミが，別の人には宝であることがある。物理学者にとっての“ゴミ”は実験データ中のバックグラウンド，つまりありふれた物理現象から生じた雑音だ。“宝”は実験データ中のシグナル。宇宙の謎を解き明かす手掛かりとなる新たな物理現象の情報だ。そして物理学の世界でも本当にゴミが宝に変わることがある。過去20年間，世界のいくつかの実験グループは陽子の崩壊現象を追い続けてきた。この現象は起こるとしても極端にまれで，そのシグナルはニュートリノというつかまえにくい素粒子のバックグラウンドに埋もれている。陽子は原子の主要な構成要素であり，永遠不滅のように思われる。だが陽子の崩壊は大統一理論の重要な予言だ。素粒子物理学ではいわゆる標準モデルが大きな成功を収めたが，大統一理論はそれをさらに発展させた有力理論だ。陽子崩壊が確認されれば，理論の正しさを示す重要な証拠になる。そこで世界各地の鉱山やトンネルの中，地下深くに，陽子崩壊を検出する巨大な実験観測装置が建設された。地下に建設するのは宇宙から降り注ぐ高エネルギー粒子「宇宙線」の影響を避けるためだ。分厚い岩盤が宇宙線を防ぐバリアの役割を果たす。ただ，このバリアは万全ではない。宇宙線が大気と衝突して生まれるニュートリノは岩盤を簡単に通り抜け，いくら地下深くに装置を建設しても，この影響を免れることはできなかった。

1980年頃から1995年頃まで世界各地で実験した第1世代の陽子崩壊実験観測装置は，陽子崩壊の兆候を観測できなかった。だが研究者たちはそれを通じて，面白みのないと思われたニュートリノのバックグラウンドが，一筋縄では説明できないことに気づいた。そうしたバックグラウンドを捉えた実験観測装置の1つが，亜鉛鉱山で知られる岐阜県神岡の山中にある「カミオカンデ」だった。名前はKamioka Nucleon Decay Experimentからとっている。超純水を詰めた大きなタンクで，その真っ暗な水中で陽子崩壊に伴うかすかな発光を高感度の検出器で捉える設計だ。

そうした事象が起こっても，干し草の山の中に隠れている1本の針のように，水分子を構成する水素や酸素の原子核とニュートリノが反応して生じた同様の1000個ほどの発光のなかに埋もれて隠されたことだろう。結局，陽子崩壊は観測できなかったが，徒労に終わったと思えたこれら1000回の発光現象の解析から正真正銘の宝物が見つかった。バックグラウンドをもたらすニュートリノは，飛んでいる間に，ある種類から別の種類に姿を変えているらしいのだ。この変身現象が本当なら，従来理論は見直しを迫られ，陽子崩壊の発見に勝るとも劣らないエキサイティングな大発見だ。

ニュートリノは何でも通り抜けてしまう驚くべき粒子だ。宇宙からは常に高速のニュートリノが降ってきていて，地球に飛来するニュートリノは毎秒，1cm^2 あたり600億個に達する。多くは太陽からやって来る。私たちの体や，その他もろもろの物体に降り注ぎ，そのまま直進して，地球の裏側に通り抜ける。これはニュートリノが他の粒子とめったに相互作用しないからだ。人体に降り注ぐ毎秒1cm^2 あたり600億個のニュートリノのうち，通常はただの1個も体を構成する原子に“かする”ことなく通り抜ける。カミオカンデのような大型装置でも，キャッチできるのは飛来する膨大な数のニュートリノのうちのごく一部だ。

標準モデルによれば，ニュートリノは3種類ある。負電荷を持つ軽い素粒子（荷電レプトン）は電子とミュー粒子，タウ粒子の3種類があり，ニュートリノと強いつながりを持つ。各ニュートリノには対応する荷電レプトンの名前が冠せられている。電子ニュートリノは原子核と衝突した際の反応で電子を生み出す。ミューニュートリノは同様の反応でミュー粒子を，タウニュートリノはタウ粒子を生み出す（ミュー粒子とタウ粒子は電子と同じ量の負電荷を持つが，ミュー粒子の質量は電子より重く，タウ粒子はさらに重い）。約70年前，原子核に関する実験結果からニュートリノの存在が予言されて以来，その質量はゼロと考えられてきた。しかし，もし種類が途中で変わるなら，量子論によれば，質量を持つことがほぼ確実になる。ニュートリノは宇宙のあらゆる場所に存在するので，質量を持つとすると，宇宙全体でのニュートリノの総質量は，すべての星を合わせた質量を上回る可能性もある。

ニュートリノを捉える巨大な罠

素粒子物理学の世界では実験に大掛かりな装置が必要で，研究を進めるにはさらに大きな装置が必要になることが多い。陽子崩壊の実験もそうだ。カミオカンデの後を継いで建設されたスーパーカミオカンデは，基本構造は同じだが大きさは10倍になっている。5万トンの超純水を満たした巨大タンクで，タンク内壁は光電子増倍管という高感度の光検出器でびっしり埋め尽くされている。

これを使って，水分子の中の陽子が崩壊するか，水を構成する原子核とニュートリノが衝突して発する微弱な光を捉える。どちらのケースでも複数の粒子が生み出され，それらが水中を高速で走る際に青白いかすかな光を進行方向前方に放つ。「チェレンコフ光」という光だ。旧ソ連の物理学者チェレンコフ（Pavel A. Cherenkov）が 1934 年に発見した現象で，超音速ジェット機などが起こす衝撃波に似ている。ジェット機が音速を超えるスピードで飛ぶと衝撃波を生じるように，荷電粒子（電子やミュー粒子など）が水中を光よりも速く移動するとチェレンコフ光が発生する。この移動は相対性理論（何ものも光速を超えて運動することはできない）には反しない。相対性理論で言う光速とは真空中の光の伝播速度のことで，水中の光の伝播速度は真空中より 25% 落ちるので超光速の粒子も存在する。

こうした荷電粒子は，スーパーカミオカンデの超純水中を数 m 走り，その間に放たれるチェレンコフ光の光の円錐がタンクの内壁に投影されて光のリングができる。このリングを光電子増倍管で捉える。リングの大きさと形状，光の強さなどからチェレンコフ光を生み出した荷電粒子の情報がわかり，それら荷電粒子を生み出したニュートリノの情報も得られる。電子が放つチェレンコフ光とミュー粒子のチェレンコフ光は簡単に区別できる。電子は高速で走るうちに多数の粒子を生み出し，それらがチェレンコフ光を発するので，タンク内壁には多くのチェレンコフ光が重なって投影されて，光のリングは輪郭がぼやける。一方，ミュー粒子は単独でチェレンコフ光を発するので，輪郭がくっきりしたリングになる。観測されたチェレンコフ光の各光電子増倍管への到達時間や，リングが内壁のどのあたりで観測されたかを調べると，もとの電子やミュー粒子のエネルギーや運動方向を知ることができる。そのデータから，ニュートリノのエネルギーや到来方向の情報が得られる。

ただ，スーパーカミオカンデは電子ニュートリノとミューニュートリノは検出できるが，タウニュートリノを捉えるのは非常に難しい。タウニュートリノと原子核の反応は，そのタウニュートリノが重いタウ粒子を生成するのに十分なエネルギーを持っている場合にのみ起こるので，反応頻度が電子ニュートリノやミューニュートリノよりも低くなるためだ。ここでいう「十分なエネルギー」とは，かなりの高エネルギーだ。ミュー粒子の質量が電子の約 200 倍なのに対し，タウ粒子の質量は電子の 3500 倍もある。タウ粒子を生み出すには，ミュー粒子を生み出す場合よりもはるかに高いエネルギーが求められる。宇宙線が大気とぶつかって生まれる大気ニュートリノは，ほとんどがミュー粒子を生み出す程度のエネルギーレベルだ。タウ粒子を生み出せるエネルギーを持つ大気ニュートリノはごくわずかなので，タウニュートリノのほとんどはエネルギー不足のため検出されないままスーパーカミオカンデを通り抜けることになる。

最も基本的な疑問は「いったいどれほどの大気ニュートリノが到来しているのか」だ。スーパーカミオカンデという素晴らしい実験観測装置が動き始めて最初の重要な仕事は「何個のニュートリノをキャッチしているか」を数えることだった。これと切っても切れないもう1つの問いは「何個くらいがキャッチされると予想されるか」だ。予想値を出すには，飛来するニュートリノが自然界でどのくらい作られているかを見積もる必要がある。

自然界には様々な起源のニュートリノが存在するが，スーパーカミオカンデが注目したのは大気ニュートリノだった。地球には宇宙線（一次宇宙線）が降り注いでいる。主に陽子で，ヘリウムや鉄など重い原子の原子核も少し混ざっている。この一次宇宙線が大気上層の原子核と衝突すると多数の粒子が生み出され，地表へ降ってくる（二次宇宙線）。その大部分はパイ中間子とミュー粒子で，これらは短時間で崩壊してニュートリノに変わる。これが大気ニュートリノだ。一次宇宙線の粒子と大気上層の原子核との衝突数（1秒間あたりの数）は大体わかっており，1回の衝突で発生するパイ中間子とミュー粒子の数も知られている。この2つのデータを組み合わせれば，毎秒，どのくらいの数の大気ニュートリノが降ってきているかが試算できる。

比を使う巧みなワザ

ただ残念なことに，こうして求めた予想値の誤差は約25％と精度がかなり低く，実測値と比べて議論するのは難しい。こうした場合によく使う手は，不確定要素が大きい2つの値の比をとることだ。比をとれば分子と分母に共通する不確定要素を取り除ける。カギを握るのは一次宇宙線が大気の原子核と反応してから最終的にニュートリノになるまでの崩壊過程だ。最初にできるパイ中間子は崩壊してミュー粒子とミューニュートリノになり，このミュー粒子はさらに崩壊して電子と電子ニュートリノ，ミューニュートリノになる。この2つの崩壊過程を合わせると，パイ中間子1個はミューニュートリノ2個と電子ニュートリノ1個，電子1個に変わる。言い換えれば，どれほど一次宇宙線の粒子が大気圏に突入してどれほどパイ中間子を生み出そうと，最終的にできるミューニュートリノと電子ニュートリノの数の比は必ず2対1になるはずだ。現実の計算は単純ではなく，大気圏に降り注ぐ一次宇宙線のシミュレーションモデルをもとに，より複雑な作業が要求される。しかし，そうして最終的に求まったミューニュートリノと電子ニュートリノの比の予想値は誤差が5%にまで収束する。

スーパーカミオカンデが観測を始めて約2年が過ぎ，かなりデータが集まった段階で，私たちはミューニュートリノと電子ニュートリノの数の比をとってみた。その結果，予想値2対1に対し，観測値は1.3対1となった。電子ニュートリノに比べミューニュートリノの数は予想値より大幅に少ない。予想と実測の大きな隔たりを埋めようとして，スーパーカミオカンデにやって来るニュートリノの推定量（フラックス）を引き上げたり，ニュートリノと超純水を構成する原子核との反応の仕方の見積もりを変えたりした。さらには，チェレンコフ光の検出感度などを再検討した。だが，観測されるミューニュートリノの少なさは説明できなかった。となると残るは，大気上層で生まれたミューニュートリノが，スーパーカミオカンデに飛んで来るまでに別の種類に“変身”した可能性だ。この驚くべき結論は正しいのだろうか。

私たちは再び，2つの値の比をとることによって，この仮説をチェックした。ある方向から来るニュートリノと，それと別の方向から来るニュートリノの数の比だ。基本的に宇宙線はどの方向からも偏りなく地球に降り注いでいるが，その一様性を乱す要素が2つある。1つは地球磁場によって一部の宇宙線粒子の飛行コースが曲げられることだ。地球磁場の影響は，速度が遅い低エネルギーの宇宙線粒子で特に大きく表れ，その到来方向に偏りが生じる。もう1つは，宇宙線の大気圏への突入角度の違いだ。天頂方向から大気圏に突入した宇宙線が生み出したパイ中間子などの粒子と，水平方向から突入した宇宙線による粒子では，通過してくる大気層の密度が違う。そのため，これら粒子の大気中での相互作用や崩壊の様子が違ってくる。この結果，垂直方向から来るニュートリノと，水平方向から来るニュートリノの観測数に違いが出る。

しかし，到来方向に関する観測数の偏りは，ある種の幾何学的な対称性を利用すれば打ち消すことができる。スーパーカミオカンデ上空から来るニュートリノと，それと 180°正反対の地球の裏側から来るニュートリノの数は，ほぼ同じになるはずだ。スーパーカミオカンデを中心に考えてみると，地球の裏側と表側の正反対の 2 方向では，同じ天頂角で大気圏に突入した宇宙線によるニュートリノが来ているからだ。地球磁場の問題については，エネルギーが高い宇宙線粒子に着目すればよい。こうした高エネルギー粒子は，地球磁場の影響をあまり受けず，もとの運動方向を保った状態で大気圏に突入して高エネルギーの大気ニュートリノを生み出す。だから正反対の 2 方向から飛来する高エネルギーのニュートリノに絞って調べれば，論理的には両者は同数になるはずだ。割り算すればきっかり 1 で，両者の比は 1 対 1 になるはずだ。ニュートリノが途中で変身して種類を変えなければ。

実際に調べてみると，高エネルギーの電子ニュートリノについては，予想通り正反対の 2 方向から来る数は同じだった。ところが高エネルギーのミューニュートリノでは，下方から来るものが上方から来るものの半分しかなかった。これが，ミューニュートリノの種類が途中で変わったことを示す第 2 の証拠だ。この結果はニュートリノの“変身”に関して，もう 1 つの重要な情報を含んでいる。下方から来たミューニュートリノは電子ニュートリノに変身したのではないということだ。もしそのように変わるのなら，その分だけ下方から到来する電子ニュートリノが増えるはずだが，実際には変わっていなかった。とすると，残る可能性はタウニュートリノしかない。ミューニュートリノがタウニュートリノに変身し，観測にかかることなく通過していったと考えられる。

化学賞
物質と生命の振る舞い

電気と物質の最新理 **Modern Theories of Electricity and Matter**

分子の実在性 **The Reality of Molecules**

ホットアトム化学 **Hot Atom Chemistry**

巨大分子はいかに作られるか **How Giant Molecules Are Made**

あるタンパク質分子の３次元構造 **The Three-Dimensional Structure of a Protein Molecule**

遺伝子抑制因子 **Genetic Repressors**

酵素機能をもつRNA **RNA as an Enzyme**

電気を通すプラスチック **Plastics That Conduct Electricity**

極微の世界をとらえるナノムービー **Filming the Invisible in 4-D**

Modern Theories of Electricity and Matter

電気と物質の最新理論（1908年掲載）

M. キュリー（1911年受賞）

マリー・キュリー（Marie Curie，1867～1934年）は現在のポーランド出身の物理学者・化学者。「ラジウムおよびポロニウムの発見とラジウムの性質およびその化合物の研究」で1911年のノーベル化学賞を単独で受賞した。また，これに先立つ1903年，「ベクレルによって発見された放射現象に関する共同研究」でノーベル物理学賞を夫のピエール・キュリー（Pierre Curie，1859～1906年）と受賞している〔同年の物理学賞の残り半分は「自発的放射能の発見」でフランスの物理学者・化学者アンリ・ベクレル（Antoine Henri Becquerel，1852～1908年）に授与〕。

ノーベル賞を2度まで受賞した女性科学者「キュリー夫人」を知らぬ人はいないだろう。まさしくスーパースターであり，女性科学者の星として多くの人々の尊敬とあこがれを集めた。

以下の記事は物理学賞の受賞から5年後，化学賞を受賞する3年前に執筆された。19世紀末から20世紀初頭は物質の原子レベルでの理解が急進展した時代であり，放射能の研究を通じてこの激変に大きく関与したキュリー自身が当時の最新の知見を解説している。抜粋記事中では言及されていないが，電子の発見は1897年，英国の物理学者ジョゼフ・ジョン・トムソン（Joseph John Thomson，1856～1940年）による。トムソンはこの業績により，1906年のノーベル物理学賞を受賞した。

初出：SCIENTIFIC AMERICAN June 1908
抜粋掲載：SCIENTIFIC AMERICAN July 2013，日経サイエンス 2013年11月号

When one reviews the progress made in the department of physics within the last ten years, he is struck by the change which has taken place in the fundamental ideas concerning the nature of electricity and matter. The change has been brought about in part by researches on the electric conductivity of gas, and in part by the discovery and study of the phenomena of radioactivity. It is, I believe, far from being finished, and we may well be sanguine of future developments. One point which appears today to be definitely settled is a view of atomic structure of electricity, which goes to conform and complete the idea that we have long held regarding the atomic structure of matter, which constitutes the basis of chemical theories.

At the same time that the existence of electric atoms, indivisible by our present means of research, appears to be established with certainty, the important properties of these atoms are also shown. The atoms of negative electricity, which we call electrons, are found to exist in a free state, independent of all material atoms, and not having any properties in common with them. In this state they possess certain dimensions in space, and are endowed with a certain inertia, which has suggested the idea of attributing to them a corresponding mass.

Experiments have shown that their dimensions are very small compared with those of material molecules, and that their mass is only a small fraction, not exceeding one one-thousandth of the mass of an atom of hydrogen. They show also that if these atoms can exist isolated, they may also exist in all ordinary matter, and may be in certain cases emitted by a substance such as a metal without its properties being changed in a manner appreciable by us.

If, then, we consider the electrons as a form of matter, we are led to put the division of them beyond atoms and to admit the existence of a kind of extremely small

Vocabulary

bring about 引き起こす
electric conductivity 電気伝導性
radioactivity 放射能
be sanguine of~ ～の自信がある
definitely 明確に
conform 一致する
complete 完成する
constitute 構成する
indivisible 分割できない
property 性質，特性
electron 電子
dimensions 大きさ
endow 付与する
inertia 慣性
attribute to (性質を)与える
molecule 分子
isolated 孤立して，単独で
appreciable はっきり感知できる，測定できる
division 区分，分類

particles, able to enter into the composition of atoms, but not necessarily by their departure involving atomic destruction. Looking at it in this light, we are led to consider every atom as a complicated structure, and this supposition is rendered probable by the complexity of the emission spectra which characterize the different atoms. We have thus a conception sufficiently exact of the atoms of negative electricity.

It is not the same for positive electricity, for a great dissimilarity appears to exist between the two electricities. Positive electricity appears always to be found in connection with material atoms, and we have no reason, thus far, to believe that they can be separated. Our knowledge relative to matter is also increased by an important fact. A new property of matter has been discovered which has received the name of radioactivity. Radioactivity is the property which the atoms of certain substances possess of shooting off particles, some of which have a mass comparable to that of the atoms themselves, while the others are the electrons. This property, which uranium and thorium possess in a slight degree, has led to the discovery of a new chemical element, radium, whose radioactivity is very great. Among the particles expelled by radium are some which are ejected with great velocity, and their expulsion is accompanied with a considerable evolution of heat. A radioactive body constitutes then a source of energy.

According to the theory which best accounts for the phenomena of radioactivity, a certain proportion of the atoms of a radioactive body is transformed in a given time, with the production of atoms of less atomic weight, and in some cases with the expulsion of electrons. This is a theory of the transmutation of elements, but differs from the dreams of the alchemists in that we declare ourselves, for the present at least, unable to induce or influence the transmutation. Certain facts go to show that radioactivity

Vocabulary

composition 構成要素, 成分
departure 離脱

supposition 推測, 仮説
render ～にする
emission spectra 放出スペクトル。spctra は spectrum の複数形

dissimilarity 違い, 相違点

in connection with ～に関連して, ～とともに

relative to ～に関する

shoot off 放出する

uranium ウラン
thorium トリウム
element 元素
radium ラジウム

velocity 速度

evolution (熱や光の)放出

account for ～を説明する

transform 変形させる, 別の物質に変える
atomic weight 原子量

transmutation of elements 元素変換
alchemist 錬金術師
induce 誘導する, 誘発する

appertains in a slight degree to all kinds of matter. It may be, therefore, that matter is far from being as unchangeable or inert as it was formerly thought and is, on the contrary, in continual transformation, although this transformation escapes our notice by its relative slowness. The conception of the existence of atoms of electricity which is thus brought before us plays an essential part in modern theories of electricity.

Vocabulary

appertain to~ （性質・属性などが）～に所属する
inert 不活性な

過去10年間の物理学分野の進歩を振り返ると，電気と物質の本質に関する基本的な考え方に生じた変化に誰もが驚かされる。この変化は，一部は気体の電気伝導性に関する研究によって，そして一部は放射能の発見と研究によってもたらされた。この変化は完了には程遠く，今後も進展が続くのは確実だろう。現段階で明確に解決したと思われるのは電気を担う原子的構造についての見方だ。この見方は化学理論の基盤をなす物質の原子構造について私たちが抱いてきた考え方と一致し，それを完成するものだ。

現在の研究手段ではそれ以上細かく分割できない“電気の原子”が存在することが確かになるとともに，それらの重要な諸性質が示された。負の電気を担うこの原子は「電子」と呼ばれ，物質の原子とは独立の自由状態に存在し，物質の原子と共通する性質は何も持っていない。自由状態の電子は空間内である大きさを占め，ある慣性を持っており，この慣性に対応する質量があると考えられる。

電子は物質の分子と比べて非常に小さいことが実験から示されており，その質量はごくわずかであって水素原子の1/1000に満たない。また，単独で存在しうるとすれば通常の物質すべてのなかにも存在している可能性があること，そして金属などの物質から，その性質に私たちが測定しうる変化を伴わない形で放出される可能性があることも示された。

もし電子を物質の一形態だと考えるなら，その区分は原子を超えたものとなり，非常に小さな粒子の存在を認めることになる。原子の構成要素となりうる小さな粒子で，しかしそれが離脱しても原子が破壊されるとは限らない粒子だ。このように考えると，どの原子も複雑な構造体だと考えられ，異なる原子を特徴づける放出スペクトルが複雑であることから，この推測はまず確実だといえる。こうして，負の電気を担う原子について十分に的確な概念が得られた。

正の電気と負の電気では非常に大きな違いがあるらしく，正の電気については事情が異なる。正の電気は常に物質の原子に付随して見られるようで，分離されうると考えてよい理由はいまのところない。また，物質に関する私たちの知識は，ある重要な事実によっても広がった。放射能と呼ばれる新しい性質が発見されたのだ。放射能はある種の物質の原子が粒子を放出する性質で，原子それ自身の質量に匹敵する粒子が放出されることもあれば，電子を放出するものもある。ウランとトリウムはわずかにこの性質を持っている。また，このように粒子を放出する性質は，非常に強い放射能を持つラジウムという新元素の発見につながった。ラジウムが放出する粒子には非常な高速で飛び出してくるものがあり，その際にかなりの熱放出を伴う。したがって放射能を持つ物体はエネルギー源となる。

放射能を説明する最良の理論によれば，放射能を持つ物質の原子のうちある割合は所定の時間内に別の物質に変化し，より原子量の小さな原子が生じ，場合によっては電子が放出される。これは元素変換だが，少なくとも現時点ではこの変換の誘発や制御は不可能なので，錬金術の夢がかなったわけではない。いくつかの事実から，どの物質もわずかな放射能を持つと考えられる。したがって，物質はこれまで考えられてきたような不変で不活性なものとは程遠く，常に変化し続けているのだろう。比較的ゆっくりした変化なので，私たちはそれに気づかないのだが。このようにして私たちにもたらされた“電気の原子”の存在という概念は，現代の電気の理論において重要な役割を担っている。

The Reality of Molecules

分子の実在性（1913年掲載）

T. スヴェドベリ（1926年受賞）

テオドール・スヴェドベリ（Theodor Svedberg，1884～1971年）はスウェーデンの化学者。1926年，「分散系に関する研究」でノーベル化学賞を単独受賞した。タンパク質などの高分子が液体中に分散したコロイド溶液の研究で優れた業績を上げた科学者だ。

以下の記事はノーベル賞受賞の13年前に書かれた。自身の専門であるコロイド溶液の研究を含め，分子・原子レベルのミクロ世界に関する理解の進展を一般向けに解説している。その意味で，マリー・キュリーによる解説記事（156ページ）と通じる面がある。

抜粋記事中に登場するラザフォード（1871～1937年）は英国の物理学者・化学者で，この記事が出版された時点ですでにノーベル賞を受賞していた（1908年化学賞）。また記事中で「ごく最近，1個の電子を分離し直接に調べた」と紹介されているミリカン（1868～1953年）は米国の物理学者で，この業績によって1923年のノーベル物理学賞を単独受賞することになる。ご存じ「ミリカンの油滴実験」だ。

原子・分子の研究に関する20世紀初めの科学界の勢いが伝わってくる。

初出：SCIENTIFIC AMERICAN February 1913
抜粋掲載：SCIENTIFIC AMERICAN July 2013, 日経サイエンス2013年11月号

Anyone consulting a handbook of chemistry or physics written toward the end of the nineteenth century, to gain information regarding molecules, would in many cases have met with rather skeptical statements as to their real existence. Some authors went so far as to deny that it would ever be possible to decide the question experimentally. And now, after one short decade, how the aspect of things is changed! The existence of molecules may today be considered as firmly established. The cause of this radical change of front must be sought in the experimental investigations of our still youthful twentieth century. [Ernest] Rutherford's brilliant investigations on α -rays, and various researches on suspensions of small particles in liquids and gases, furnish the experimental substantiation of the atomistic conception of matter.

The modern proof for the existence of molecules is based in part upon phenomena which give us a direct insight into the discontinuous (discrete) structure of matter, and in part upon the "working model" of the kinetic theory furnished us in colloidal solutions. These last have been shown to differ from "true" solutions only in that the particles of the dissolved substance are very much larger in the case of colloids. In all respects they behave like true solutions, and follow the same laws as the latter. And, thirdly, the recent direct proof of the existence of indivisible elementary electric charges enable us to draw conclusions regarding the atomic structure of ponderable matter.

Among the first-mentioned class of proofs is Rutherford's great discovery (1902–1909) that many radioactive substances emit small particles which, after losing their velocity, as for instance by impact against

Vocabulary

molecule 分子
skeptical 懐疑的な
go so far as to do ～さえする，～するほどまでである

change of front 見解の変化，態度の変化
sought seek（理由・説明などを求める）の過去分詞
α -ray アルファ線
suspension 浮遊

substantiation 証拠，実証

discontinuous 不連続な
discrete 離散的な
working model 作業モデル
kinetic 動力学の，動的な
colloidal solution コロイド溶液
▶ Technical Terms

indivisible 不可分な

radioactive 放射性の
emit 放出する

Technical Terms

コロイド溶液（**colloidal solution**）　微細な粒子や液滴が別種の液体中に分散したもの。身の回りに見られる例では牛乳など。

the walls of a containing vessel, display the properties of helium gas. In this way it has been proved experimentally that helium is built up of small discrete particles, molecules. In fact, Rutherford was able actually to count the number of α particles or helium molecules contained in one cubic centimeter of helium gas at 0 degree Centigrade and one atmosphere pressure (1908).

The second class of proofs of the existence of molecules comprises a number of researches on the change of concentration with level which is observed in colloidal suspensions, and on the related phenomena of diffusion, Brownian movement, and light absorption in such systems.

Lastly, modern investigations of the conduction of electricity through gases, and of the so-called β-rays, have shown conclusively that electric charges, like matter, are of atomic nature, i.e., composed of ultimate elementary charged particles, whose mass is only about 1/700 of a hydrogen atom. Quite recently [Robert Andrews] Millikan and [Erich] Regener have succeeded by entirely different methods in isolating an electron and studying it directly.

We see, then, that the scientific work of the past decade has brought most convincing proof of the existence of molecules. Not only is the atomic structure of matter demonstrated beyond reasonable doubt, but means have actually been found to study an individual atom. We can now directly count and weigh the atoms. What skeptic could ask for more?

Vocabulary

helium ヘリウム

α particle アルファ粒子

comprise ～からなる
concentration 濃度
colloidal suspension コロイド懸濁液
diffusion 拡散
Brownian movement ブラウン運動

β -ray ベータ線
▶ Technical Terms

convincing 説得力のある, 確信を抱かせる

reasonable doubt 合理的疑い

Technical Terms

ベータ線(**β -ray**) 原子核のベータ崩壊に伴って発せられる放射線で, 電子線のこと(広義には陽電子も含む)。ちなみにアルファ線(**α -ray**)はヘリウムの原子核であるアルファ粒子の放射であり, ガンマ線(**γ -rays**)は波長の極めて短い電磁波だ。

分子について調べようと19世紀末に書かれた化学や物理のハンドブックを見ると，多くの場合，分子の実在についてかなり懐疑的な記述にぶつかるだろう。分子の存在を実験で決定づけるのは永遠に不可能だろうと述べている例さえある。それからわずか10年しかたっていないのに，事態はいかに変わったことか！　現在，分子の存在は確固たるものだと考えられている。この見解の急変の原因は，まだ始まったばかりの20世紀の実験研究に求められねばならない。ラザフォード（Ernest Rutherford）によるアルファ線に関する素晴らしい研究と，液体や気体中の小さな浮遊粒子に関する様々な研究が，物質を原子レベルでとらえる考え方に実験的な根拠を与えている。

分子の存在についての最新の証明は，物質の不連続（離散的）な構造を直接うかがわせる諸現象と，コロイド溶液の動力学に関する“作業モデル”に基づいている。コロイド溶液と“真の”溶液との違いは，溶けている物質の粒子の大きさがコロイド溶液ではずっと大きいという点だけであることが示された。コロイド溶液はすべての面で真の溶液と同様に振る舞い，同じ法則に従う。そして第3に，電荷を担う不可分の基本粒子の存在が最近になって直接的に証明されたことで，物質の原子レベルの構造に関する結論を導けるようになった。

最初に触れた部類の証明の一例は，ラザフォードによる素晴らしい発見（1902～1909年）だ。多くの放射性物質が小さな粒子を放出しており，その粒子が容器の壁にぶつかるなどして速度を失うとヘリウムガスの性質を示すことを彼は発見した。こうして，ヘリウムが個別の小さな粒子，つまり分子からできていることが実験的に証明されたのだ。実際，ラザフォードは0℃・1気圧のヘリウムガス$1cm^3$に含まれるアルファ粒子すなわちヘリウム分子の数を正確に計測することに成功した（1908年）。

2番目の部類の証明は，コロイド懸濁液に見られる深さ方向の濃度変化や，関連の拡散やブラウン運動，光の吸収に関するいくつかの研究からなる。

そして最後に，ガス中の電気伝導と，いわゆるベータ線を調べた最新の研究から，電荷に物質と同様の原子的な性質があること，つまり質量が水素原子の約1/700にすぎない究極の荷電素粒子の存在が示された。ごく最近，ミリカン（Robert Andrews Millikan）とレーゲナー（Erich Regener）がそれぞれまったく異なる方法を用いて1個の電子を分離し，それを直接に調べた。

このように，過去10年の科学研究が分子の存在について極めて説得力の大きな証明をもたらしたことがわかる。物質の原子構造が合理的疑いの余地なく示されただけでなく，個別の原子を調べる方法が実際に見つかったのだ。私たちは現在，原子の数を数え，その重さを量ることができる。これでも疑わしい点がまだ何か残っているだろうか？

Hot Atom Chemistry

ホットアトム化学（1950年掲載）

W. F. リビー（1960年受賞）

ウィラード・フランク・リビー（Willard Frank Libby, 1908～1980年）は米国の化学者。1960年に「炭素14による年代測定法の研究」によってノーベル化学賞を単独受賞した。授賞理由となった業績が最も有名だが，核化学・放射化学分野で様々な研究に取り組み，第二次世界大戦中はマンハッタン計画に参加してウラン濃縮技術（気体拡散法）の開発に大きく寄与した。

ノーベル賞受賞の10年前に執筆された以下の記事は炭素同位体年代測定法そのものを述べたものではなく，「ホットアトム化学」という研究分野を解説している。原子核の崩壊などで生じた新たな原子は非常に高いエネルギーに励起されており，ホットアトムと呼ばれる。こうした高エネルギーの原子が引き起こす化学反応を研究するのがホットアトム化学だ。

ビーカーとフラスコでできる実験ではないので，その意味では特殊といえるが，同位体の分離濃縮や放射性の標識化合物の製造などに利用されている。

初出：SCIENTIFIC AMERICAN March 1950
抜粋掲載：SCIENTIFIC AMERICAN July 2013, 日経サイエンス 2013年11月号

One of the first things a beginning chemistry student learns is that the chemical behavior of an atom depends solely on the electrons circulating around the nucleus, and not at all on the nucleus itself. In fact, the classical definition of isotopes states that all the isotopes of a given element are identical in chemical activity, even though the nuclei are different. Like all generalizations, even this one has a little bit of falsehood in it. The truth is that the chemical behavior of an atom may be strongly influenced by events in its nucleus, if the nucleus is radioactive. The bizarre chemical effects sometimes produced by radioactive atoms have given rise to a fascinating new branch of investigation known as hot atom chemistry.

Unusual chemical reactions among hot atoms were noticed soon after the discovery of radioactivity. The serious study of hot atom chemistry began as early as 1934, when Leo Szilard and T. A. Chalmers in England devised a method, known as the Szilard-Chalmers process, for utilizing such reactions to obtain concentrated samples of certain radioactive compounds for research purposes. But not until the end of the recent war, when chemists began to work with large amounts of radioactive materials, did the subject begin to attract wide interest. Since the war, reports of investigations in this intriguing field have come from laboratories in all the leading scientific countries of the world.

The particular set of reactions we shall consider is the behavior of radioactive iodine in the compound ethyl iodide—CH_3CH_2I. We begin with an ordinary liq-

Vocabulary

nucleus 原子核

isotope 同位体
element 元素
identical 同一である

generalization 一般化
falsehood 誤り，偽り

radioactive 放射性の
bizarre 奇怪な，一風変わった

hot atom ホットアトム
▶ Technical Terms

radioactivity 放射能

Szilard-Chalmers process シラード・チャルマース反応
utilize 利用する

intriguing 興味深い

iodine ヨウ素
ethyl iodide ヨウ化エチル

Technical Terms

ホットアトム（**hot atom**） 放射性の原子核が崩壊したり，原子核に外部から粒子をぶつけて別種の原子核に変換したりすると，一般に大きな運動エネルギーや電荷を獲得して高いエネルギー状態になる。こうして著しい励起状態になった原子がホットアトムで，ホットアトムが引き起こす特異な化学反応を調べる分野をホットアトム化学と呼んでいる。ホットアトムの研究は本文で続いて紹介されている「シラード・チャルマース反応」の発見が出発点となった。

uid sample of the compound and transform some of the iodine atoms in it into a radioactive variety by irradiating them with neutrons from a chain-reacting pile or a cyclotron. Neutrons have no chemical properties, since they consist of pure nuclear matter with no associated external electrons. Because they have no external electrons, and are themselves electrically neutral, their penetrating power is amazing. They readily proceed through several inches of solid material until they chance to interact with some of the tiny atomic nuclei in their path.

Suppose, then, we expose a bottle of liquid ethyl iodide to a source of neutrons. The neutrons penetrate the glass, and a certain proportion of them are captured by the iodine atoms. When the nucleus of a normal iodine atom, I-127, takes in a neutron, it is transformed into the radioactive isotope I-128. This new species is extremely unstable: in much less than a millionth of a millionth of a second it emits a gamma ray of huge energy—several million electron volts. After giving off this tremendous energy, the I-128 atom is reduced to a lower state of excitation. It is still unstable; the atom continues to decay, and gradually, with a half-life of 25 minutes, the I-128 atoms degenerate into xenon 128 by emitting beta particles. The emission of this energy gives the I-128 atom in the ethyl-iodide molecule a large recoil energy, just as the firing of a bullet from a gun makes the gun recoil. The atom's recoil energy is calculated to be some 200 million electron volts. Now the chemical energy with which the iodine atom is bound in the ethyl-iodide molecule is only about three or four electron volts. The energy of recoil is so much greater than the strength of

Vocabulary

irradiate 照射する
neutron 中性子
pile 原子炉
cyclotron サイクロトロン
▶ Technical Terms
external electron 外部電子
penetrate 透過する
interact 相互作用する
capture 捕獲する
take in 取り込む
transforme （別のものに）変える
emit 放出する
gamma ray ガンマ線
electron volt 電子ボルト（エネルギーの単位）
state of excitation 励起状態
decay 崩壊する
half-life 半減期
xenon キセノン
beta particle ベータ粒子（電子）
recoil energy 反跳エネルギー
bound bind（結びつける，束縛する）の過去分詞

Technical Terms

サイクロトロン（**cyclotron**）　電場と磁場を用いて粒子を加速する円形加速器の一種で，中心部の粒子源から出た粒子が螺旋状の軌道を描きながら加速される。1930年に米国の物理学者ローレンスらが発明した（82ページ参照）。後に，磁場と電場の周波数を粒子の加速に合わせて制御することで粒子を円軌道で加速するシンクロトロン（**synchrotron**）に発展した。現在主流となっている円形加速器はシンクロトロンだ。

the chemical bond that every I-128 atom is ejected from its molecule with considerable force. Hot atom chemistry is concerned with the unusual chemical reactions that these high-velocity iodine atoms undergo after they are expelled from the molecule. Since the 1-128 atoms are radioactive, it is relatively easy to trace them through their subsequent activities.

To what uses can hot atom chemistry be put? One of the obvious uses is the preparation of extremely concentrated sources of radioactivity. This technique should be of assistance in many purposes for which radioactive material is used, notably in biology. When a radioactive isotope is injected into the body, either as a tracer or in a treatment for disease, it is often essential that the amount of material injected be held to a minimum, in order to avoid disturbance of the normal constitution of the blood or the normal metabolism of the body.

Vocabulary

chemical bond 化学結合
considerable かなりの
expell 放出する

trace 追跡する
subsequent その後の

put A use to~ Aを～に利用する

tracer トレーサー, 追跡用の目印

constitution 組成
metabolism 代謝

化学を学び始めた学生が最初に教わることのひとつに，原子の化学的振る舞いは原子核を周回している電子のみによって決まり，原子核自体にはよらないということがある。実際，元素の同位体は原子核の構成が異なるにもかかわらず化学的な活動はみな同じである，というのが同位体の古典的な説明となっている。しかしあらゆる一般化と同様，この一般化にも小さな誤りがある。原子核が放射性である場合，その原子の化学的振る舞いが原子核内部の出来事に強く影響される場合があるのだ。放射性原子が時に引き起こすこの変わった化学効果は，「ホットアトム化学」という魅力的な新研究分野を生んだ。

ホットアトムが示す異常な化学反応は放射能が発見されて間もなく気づかれた。ホットアトム化学の研究が本格化したのは1934年，英国のシラード（Leo Szilard）とチャルマース（T. A. Chalmers）が「シラード・チャルマース反応」を考案したときにさかのぼる。ある種の放射性化合物の高濃度サンプルを研究用に得るために考案された反応だ。しかし，広く関心を集めるようになったのは最近で，第二次世界大戦末期に化学者が大量の放射性物質を扱うようになってからだった。戦後は，この興味深い領域に関する研究報告が世界中の科学先進国から生まれている。

ここではヨウ化エチル CH_3CH_2I に含まれる放射性ヨウ素の振る舞いについて考えよう。まず通常のヨウ化エチル試料液を用意し，連鎖反応中の原子炉またはサイクロトロンからの中性子を照射することで，ヨウ化エチル中のヨウ素原子の一部を放射性のものに変える。中性子は電荷を持たず電子を伴わないから，化学的特性はない。外部電子を持たないうえ，それ自身は電気的に中性だから，中性子は驚異的な透過力を発揮する。固体物質中をやすやすと数インチ進み，経路沿いにある小さな原子核のいくつかとたまたま相互作用してようやく止まる。

だから，ヨウ化エチル液をガラス瓶に入れて中性子源にさらすと，中性子はガラスを貫通した後，一部がヨウ素原子に捕獲される。通常のヨウ素127の原子核が1個の中性子を取り込むと，放射性同位体のヨウ素128に変わる。この新しい核種は非常に不安定で，100万分の1秒の100万分の1もたたないうちに高エネルギー（数百万電子ボルト）のガンマ線を放出する。この大きなエネルギーを放出して，ヨウ素128原子は低い励起準位に落ちるが，それでもまだ不安定だ。さらに崩壊を続け，ヨウ素128原子はベータ粒子を放出して半減期25分でキセノン128に変わる。このベータ粒子放出の際に，ヨウ化エチル中のヨウ素128原子は大きな反跳エネルギーを得る。銃が弾丸を発射する際に反動するのと同様だ。反跳エネルギーは2億電子ボルトほどになる計算で，これに対しこのヨウ素原子をヨウ化エチル中に結びつけている化学エネルギーはたったの3～4電子ボルトにすぎない。化学結合のエネルギーよりも反跳エネルギーのほうがはるかに大きいので，すべてのヨウ素128原子が分子からかなりの力で飛び出す。こうして分子から放出された高速のヨウ素原子がその後に引き起こす特殊な化学反応を扱うのが，ホットアトム化学だ。ヨウ素128の原子は放射性なので，その後の活動を追跡するのは比較的容易である。

ホットアトム化学は何に利用できるのか？　明らかな用途に，非常な高濃度の放射線源の作成がある。この技術は放射性物質を用いる様々な用途，特に生物学に役立つはずだ。トレーサーとして，あるいは治療目的で放射性同位体を人体に注入する場合には，血液組成や人体の代謝が乱されるのを避けるため，注入量を最小限に抑えることが不可欠だ。

How Giant Molecules Are Made

巨大分子はいかに作られるか（1957年掲載）

G. ナッタ（1963年受賞）

ジュリオ・ナッタ（Giulio Natta, 1903～1979年）はイタリアの化学者。1963年，「新しい触媒を用いた重合法の発見とその基礎的研究」によって，ドイツの化学者カール・ツィーグラー（Karl Ziegler，1898～1973年）とともにノーベル化学賞を受賞した。エチレンなどの重合に使われるいわゆる「ツィーグラー・ナッタ触媒」の開発者だ。

エチレンを重合してポリエチレンを作るにはかつては高い圧力が必要だったが，ツィーグラーはこれを常圧で実現する触媒を1953年に開発した。ナッタはこれにさらに改良を加え，プロピレンの重合を実現した。20世紀は軽くて強く安価なプラスチックが普及して，人々の暮らしを大きく変えた世紀だった。合成樹脂を効率よく工業的に生産する要がこうした触媒であり，その恩恵は計り知れない。

ノーベル賞受賞の6年前に執筆された以下の記事は，1950年代における触媒の研究開発と石油化学産業の姿を伝えている。高分子化学の進歩と実用技術の開発，それをいち早く工業化する産業界の意欲が明らかだ。

初出：SCIENTIFIC AMERICAN September 1957
抜粋掲載：SCIENTIFIC AMERICAN July 2013, 日経サイエンス 2013年11月号

A chemist setting out to build a giant molecule is in the same position as an architect designing a building. He has a number of building blocks of certain shapes and sizes, and his task is to put them together in a structure to serve a particular purpose. The chemist works under the awkward handicap that his building blocks are invisible, because they are submicroscopically small, but on the other hand he enjoys the happy advantage that nature has provided models to guide him. By studying the giant molecules made by living organisms, chemists have learned to construct molecules like them. What makes high-polymer chemistry still more exciting just now is that almost overnight, within the last few years, there have come discoveries of new ways to put the building blocks together—discoveries which promise a great harvest of new materials that have never existed on the earth.

We can hardly begin to conceive how profoundly this new chemistry will affect man's life. Giant molecules occupy a very large place in our material economy. Tens of millions of men and women, and immense areas of the earth's surface, are devoted to production of natural high polymers, such as cellulose, rubber and wool. Now it appears that synthetic materials of equivalent or perhaps even better properties can be made rapidly and economically from coal or petroleum. Among other things, this holds forth the prospect that we shall be able to turn much of the land now used for the production of fiber to the production of food for the world's growing population.

Free radicals are one type of catalyst that can grow polymers by addition; another method involves the use of ions as catalysts. The latter is a very recent development, and to my mind it portends a revolution in the synthesis of giant molecules, opening up large new horizons. The cationic method has produced some very interesting high polymers: for instance, butyl rubber, the

Vocabulary

giant molecule 巨大分子

awkward 厄介な, やりにくい, 困った

submicroscopically 顕微鏡でも見えないほどに

high-polymer chemistry 高分子化学

cellulose セルロース

equivalent 同等の

hold forth 提示する

prospect 予想, 見通し

free radical 遊離基, フリーラジカル

catalyst 触媒

addition 付加反応

portend ～の前触れとなる

cationic method カチオン法

butyl rubber ブチルゴム

synthetic rubber used for tire inner tubes. But the anionic catalysts, a more recent development, have proved far more powerful. They yield huge, made-to-order molecules with extraordinary properties.

Early in 1954 our group in the Institute of Industrial Chemistry of the Polytechnic Institute of Milan, using certain special catalysts, succeeded in polymerizing complex monomers of the vinyl family. We were able to generate chains of very great length, running to molecular weights in the millions (up to 10 million in one case). We found that it was possible, by a proper choice of catalysts, to control the growth of chains according to predetermined specifications.

Among the monomers we have polymerized in this way are styrene and propylene, both hydrocarbons derived from petroleum. The polypropylenes we have made illustrate the versatility of the method. We can synthesize them in three forms: isotactic, atactic or "block isotactic," that is, a chain consisting of blocks, one having all the side groups aligned on one side, the other on the opposite side. The isotactic polypropylene is a highly crystalline substance with a high melting point (346 degrees Fahrenheit); it makes very strong fibers, like those of natural silk or nylon. The atactic product, in contrast, is amorphous and has the elastic properties of rubber. The block versions of polypropylene have the intermediate characteristics of a plastic, with more or less rigidity or elasticity.

The possibility of obtaining such a wide array of different products from the same raw material

Vocabulary

anionic catalyst アニオン触媒
made-to-order 特注の

polymerizie 重合させる
monomer モノマー

molecular weight 分子量

specifications スペック，仕様

styrene スチレン
propylene プロピレン
hydrocarbon 炭化水素
versatility 汎用性

isotactic アイソタクチック
▶ Technical Terms
atactic アタクチック
▶ Technical Terms

crystalline 結晶性の
melting point 融点

amorphous 非晶質，アモルファス，
elastic 弾性のある
intermediate 中間の
rigidity 剛性
elasticity 弾性

Technical Terms

アイソタクチック（**isotactic**）　鎖状の炭化水素高分子などに生じる立体異性の一種で，主鎖に結びついた置換基のうち問題にしている置換基が主鎖の同じ側に並んでいるもののこと。
アタクチック（**atactic**）　同じく，問題にしている置換基が主鎖に対してランダムに配置しているもののこと。

naturally aroused great interest. Furthermore, the new controlled processes created properties not attainable before: for example, polystyrene, which had been known only as a glassy material with a low softening point (under 200 degrees F), now could be prepared as a strong, crystalline plastic with a melting point near 460 degrees. The new-found power of the anionic catalysts stimulated great activity in polymer research, both in Europe and in the U.S. New polymers were made from various monomers. In our own laboratory we synthesized all of the regular polymers, and some amorphous ones, that can be made from butadiene; some of the products are rubber-like, others not. At about the same time the B. F. Goodrich Company and the Firestone Tire and Rubber Company both announced that they had synthesized, from isoprene, a rubber identical to natural rubber—a problem on which chemists throughout the world had worked in vain for more than half a century.

In some respects we can improve on nature. As I have mentioned, we shall probably be able to create many new molecules which do not exist in living matter. They can be made from simple, inexpensive materials. And we can manufacture giant molecules more rapidly than an organism usually does. Although it is less than four years since the new methods for controlled synthesis of macromolecules were discovered, already many new synthetic substances—potential fibers, rubbers and plastics—have been made.

Vocabulary

attainable 達成可能な

softening point 軟化点
▶ Technical Terms

butadiene ブタジエン

isoprene イソプレン
natural rubber 天然ゴム
in vain いたずらに，むなしく

Technical Terms

軟化点（**softening point**） プラスチックやガラスなどの物質を加熱したとき，軟化して変形し始める温度のこと。これらの物質には明確な融点を持たないものもあり，加熱に伴ってしだいに溶融状態になるので，融点の代わりにこう呼んでいる。

巨大分子を作り上げようとしている化学者は，ビルを設計している建築家と同じ立場にいる。ある形と大きさの基礎材料をいくつか手にし，それらを組み合わせて特定の目的にかなう構造物を作るのが仕事だ。化学者の場合，この基礎材料が小さくて顕微鏡でも見えないという厄介なハンデを負っているものの，一方では自然が手引きとなるモデルを提供してくれるという，うれしい強みがある。化学者は生物が作り出す巨大分子を研究することで，同様の分子の構築法を学んできた。現在の高分子化学をさらにエキサイティングにしているのは，そうした基礎材料を組み合わせる新たな方法が過去 2 ～ 3 年で相次いで発見されたことだ。地球上にかつて存在したことのない新素材を大量に収穫できるようになる。

この新しい化学が人間の生活に与える影響の大きさは計り知れない。巨大分子は私たちの物質経済に非常に大きな位置を占めている。何千万人もの人々と地表の莫大な領域が，セルロースやゴム，羊毛など天然の高分子の生産に充てられている。それがいま，それらと同等あるいはより優れた特性の合成材料を，石炭や石油から素早く経済的に作れるようになった。現在は繊維の生産に使われている土地の多くを，増え続ける世界人口を養う食物の生産に振り向けることも可能になるだろう。

付加反応によって高分子を成長させる触媒にフリーラジカルがあるが，別の方法ではイオンを触媒として用いる。これはごく最近に開発された方法であり，今後の巨大分子の合成に革命をもたらして大きな新地平を開くと思う。カチオン法はすでに，タイヤチューブに使われる合成ゴムであるブチルゴムなど，非常に興味深い高分子をいくつか作り出した。より最近に開発されたアニオン触媒はさらに強力で，並外れた特性を持つオーダーメードの巨大分子を生み出している。

私たちミラノ工科大学工業化学研究所のチームは 1954 年初め，特殊な触媒を用いて複雑なビニル系モノマーの重合に成功した。非常に長いポリマー鎖，分子量が数百万（ある例では 1000 万）に達するものを合成できた。触媒を適切に選ぶことで，事前の計画通りにポリマー鎖の成長を制御できることがわかった。

そのように重合できたモノマーにスチレンとプロピレンがある。どちらも石油から誘導された炭化水素だ。私たちが作ったポリプロピレンはこの方法の汎用性を如実に示している。アイソタクチックとアタクチック，そして“ブロック・アイソタクチック”という3つの形態のポリプロピレンを合成できた。ブロック・アイソタクチックとは，すべての側基が主鎖の片側にそろって並んだブロックと，反対側に並んだブロックとが，交互につながったポリマーだ。アイソタクチック型のポリプロピレンは高い融点（174℃）を持つ結晶性の物質で，絹やナイロンと同様の高強度の繊維になる。アタクチック型は対照的にアモルファス（非晶質）で，ゴムのような弾性を持つ。ブロック型ポリプロピレンはこれらの中間的な特性を持つ。

同じ原材料からこのように様々な製品が得られるということで，当然ながら非常に大きな関心が集まった。さらに，この新制御プロセスによってこれまでにない特性が作り出された。例えばポリスチレンはこれまで低い軟化点（90℃以下）のガラス質の素材しか知られていなかったが，現在は融点が240℃近い強靭な結晶性プラスチックにできるようになっている。アニオン触媒に見つかったこの新たな力は，ヨーロッパと米国の両方で高分子研究を大きく刺激し，様々なモノマーから新しいポリマーが作られている。私たちの研究室はブタジエンから通常のポリマー全タイプと，アモルファスなポリマーをいくつか合成した。あるものはゴムのようで，他はそうではない。これとほぼ同時に，タイヤメーカーのグッドリッチとファイアストンがそれぞれ，イソプレンから天然ゴムとまったく同じゴムを合成したと発表した。世界中の化学者が半世紀以上も取り組んできたのに実現できていなかった偉業だ。

ある意味で，私たちは自然を改良できる。この記事で述べたように，生体中には存在しない様々な新分子を作り出すことが可能だろう。それらは単純で安価な物質から作れる。そして生物が行っているよりも迅速に巨大分子を製造できる。巨大分子を調節しつつ合成するこれらの新手法が発見されてからまだ4年に満たないが，すでに繊維やゴム，プラスチックなど多数の合成素材ができている。

The Three-Dimensional Structure of a Protein Molecule

あるタンパク質分子の３次元構造（1961年掲載）

J. C. ケンドリュー（1962年受賞）

ジョン・ケンドリュー（John Cowdery Kendrew，1917～1997年）は英国の生化学者・結晶学者。1962年，「球状タンパク質の構造研究」によって英国の化学者マックス・ペルーツ（Max Ferdinand Perutz，1914～2002年）とともにノーベル化学賞を受賞した。2人は英ケンブリッジ大学キャベンディッシュ研究所の同僚。赤血球にあるヘモグロビンを結晶化してX線回折像を撮影することで，このタンパク質分子の立体構造を明らかにした。タンパク質構造解析の草分けだ。

ノーベル賞受賞の前年に執筆された以下の記事は，2人が取り組んだ実験を具体的に述べている。これまで誰も目にしたことのないタンパク質分子の姿を世界で初めて目にする興奮が率直につづられており，興味深い。

記事にも触れられているように，タンパク質分子の立体構造解析はタンパク質の生化学的な機能を理解するうえで欠かせない。その後半世紀，主にコンピューター技術の進展によってX線回折データの処理は格段に進歩し，高輝度のX線源も実現している。

初出：SCIENTIFIC AMERICAN December 1961
抜粋掲載：SCIENTIFIC AMERICAN July 2013, 日経サイエンス 2013年11月号

When the early explorers of America made their first landfall, they had the unforgettable experience of glimpsing a New World that no European had seen before them. Moments such as this— first visions of new worlds—are one of the main attractions of exploration. From time to time scientists are privileged to share excitements of the same kind. Such a moment arrived for my colleagues and me one Sunday morning in 1957, when we looked at something no one before us had seen: a three-dimensional picture of a protein molecule in all its complexity. This first picture was a crude one, and two years later we had an almost equally exciting experience, extending over many days that were spent feeding data to a fast computing machine, of building up by degrees a far sharper picture of this same molecule. The protein was myoglobin, and our new picture was sharp enough to enable us to deduce the actual arrangement in space of nearly all of its 2,600 atoms. We had chosen myoglobin for our first attempt because, complex though it is, it is one of the smallest and presumably the simplest of protein molecules, some of which are 10 or even 100 times larger.

In a real sense, proteins are the "works" of living cells. Almost all chemical reactions that take place in cells are catalyzed by enzymes, and all known enzymes are proteins; an individual cell contains perhaps 1,000 different kinds of enzyme, each catalyzing a different and specific reaction. Proteins have many other important functions, being constituents of bone, muscle and tendon, of blood, of hair and skin and membranes. In addition to all this it is now evident that the hereditary information, transmitted from generation to generation in the nucleic acid of the chromosomes, finds its expression in the characteristic types of protein molecule synthesized by each cell. Clearly to understand the behavior of a living cell it is necessary first to find out how so wide a variety of functions can be assumed by molecules all made up for the most part of the same few basic units.

Vocabulary

make landfall 上陸する
glimpse 垣間見る
privilege 特権を与える
protein タンパク質
crude 粗雑な
myoglobin ミオグロビン
catalyze 触媒する, 触媒作用によって反応を促進する
enzyme 酵素
constituent 成分, 構成物
tendon 腱
nucleic acid 核酸
chromosome 染色体
assume 役目を引き受ける

These units are amino acids, about 20 in number, joined together to form the chains known as polypeptides. The hemoglobin in red blood corpuscles contains four polypeptide chains. Myoglobin is a junior relative of hemoglobin, consisting of a single polypeptide chain.

Even in the present incomplete state of our studies on myoglobin we are beginning to think of a protein molecule in terms of its three-dimensional chemical structure and hence to find rational explanations for its chemical behavior and physiological function, to understand its affinities with related proteins and to glimpse the problems involved in explaining the synthesis of proteins in living organisms and the nature of the malfunctions resulting from errors in this process. It is evident that today students of the living organism do indeed stand on the threshold of a new world. Analyses of many other proteins, and at still higher resolutions (such as we hope soon to achieve with myoglobin), will be needed before this new world can be fully invaded, and the manifold interactions between the giant molecules of living cells must be comprehended in terms of well-understood concepts of chemistry.

Nevertheless, the prospect of establishing a firm basis for an understanding of the enormous complexities of structure, of biogenesis and of function of living organisms in health and disease is now distinctly in view.

Vocabulary

amino acid アミノ酸
polypeptide ポリペプチド
▶ Technical Terms
hemoglobin ヘモグロビン
red blood corpuscle 赤血球。red blood cell に同じ
rational 理にかなった
affinity 親和性。ここでは「関係」。
synthesis 合成
malfunction 機能不全，動作不良
resolution 解像度
invade 分け入る
manifold 多種多様な
biogenesis 生物発生

Technical Terms

ポリペプチド(**polypeptide**)　多数のアミノ酸がつながった化合物。タンパク質とほぼ同義だが，タンパク質よりも小さな分子を指すことが多い。はっきりした境界線はないので，小ぶりのタンパク質と考えて差し支えない。

かつて探検家たちがアメリカ大陸に初めて上陸したとき，彼らはヨーロッパ人がそれまで見たことのない新世界を目にするという忘れがたい経験をした。新世界を最初に目撃するそうした瞬間が，探検の醍醐味だ。科学者も時折，同種の興奮を味わう機会に恵まれる。私と共同研究者にそうした瞬間が訪れたのは 1957 年のある日曜日の朝，ある複雑なタンパク質分子の初めての 3 次元写真を見たときだった。この最初の写真は粗いものだったが，その 2 年後，何日もかけて高速計算機にデータを入力して同じ分子のはるかに鮮明な画像を少しずつ作り上げるという，同じくらいエキサイティングな経験をした。このタンパク質はミオグロビンと呼ばれるもので，新しい写真は十分に鮮明で，その 2600 個の原子のほぼすべての空間的配置を推定できた。最初の撮影対象にミオグロビンを選んだのは，複雑な分子ではあるが，タンパク質としては最も小さい部類で単純だと考えられたからだ。タンパク質にはミオグロビンよりも 10 倍，あるいは 100 倍大きなものもある。

実際的な意味で，タンパク質は細胞のなかの“工場”だ。細胞中で起きている化学反応のほとんどは酵素によって触媒されており，知られる酵素はすべてタンパク質である。個々の細胞に含まれる酵素はおそらく 1000 種類に上り，それぞれが特定の反応を触媒している。タンパク質はこのほかにも多くの重要な機能を果たしており，骨や筋肉，腱，血液，毛髪，皮膚，膜組織の成分となっている。加えて，染色体中の核酸として世代から世代へ伝えられている遺伝情報が，各細胞が合成するタンパク質分子の特徴的タイプとして発現していることが，いまや明らかになっている。生きている細胞の振る舞いを明確に理解するには，まず，少数の同じ基本ユニットからできあがっているタンパク質分子がどのようにしてこれほど多様な機能を発揮するのかを解明する必要がある。

それらのユニットは約 20 種類のアミノ酸であり，いくつかが集まってポリペプチドという鎖を形成する。赤血球にあるヘモグロビンは 4 つのポリペプチド鎖を含んでいる。ミオグロビンはヘモグロビンの小型版で，ポリペプチド鎖 1 つだけでできている。

ミオグロビンに関する私たちの研究はまだ不完全だが，それでも私たちは 1 つのタンパク質分子を 3 次元の化学構造としてとらえることによってタンパク質分子の化学的振る舞いと生理的機能を合理的に説明して，他の関連タンパク質との関係や，生体におけるタンパク質合成とその過程のエラーから生じる機能不全の本質に関わる問題の解明に取り組み始めている。現在，生物を研究している者がまさに新世界の入り口に立っているのは明らかだ。この新世界に深く分け入るには，他の多くのタンパク質をさらに高解像度で解析することが必要だろう（ミオグロビンについては近くそうした高解像度の解析ができるようになると私たちは期待している）。そして，生きている細胞のなかでタンパク質という巨大分子の間に働いている多種多様な相互作用を，十分に理解されている化学の考え方に基づいて理解しなければならない。

だが，生物の構造や生物発生，健康な生体や病気の生体における機能など，とてつもなく複雑な事柄を理解するためのしっかりした基礎を確立できる見込みが，いまやはっきりと視野に入っている。

Genetic Repressors

遺伝子抑制因子（1970年掲載）

M. プタシュネ／ W. ギルバート（1980年受賞）

ウォルター・ギルバート（Walter Gilbert，1932年～）は米国の物理学者・生化学者。1980年，「核酸の塩基配列の決定」によって，英国の生化学者フレデリック・サンガー（Frederick Sanger，1918～2013年）とともにノーベル化学賞を受賞した。授賞業績はDNAの塩基配列を調べる化学的な解析法の開発で，ギルバートの方法は化学分解法，サンガーのものは酵素法と呼ばれる。ちなみに同年の化学賞の残り半分は「遺伝子工学の基礎としての核酸の生化学的研究」によって米国の生化学者ポール・バーグ（Paul Berg，1926年～）に授与された。

以下の記事はノーベル賞受賞の10年前に執筆された。塩基配列決定法の説明ではなく，遺伝子がオン・オフされるメカニズムについて，1970年当時の最新の知見を概説している。当時は分子生物学が著しい技術的進展を始めたころで，抜粋記事にあるように遺伝子制御に関わる様々な因子が同定され，その機能が明らかになっていった。生化学的な実験によって生命のメカニズムに分子レベルで迫ることが可能になった。

ただ，遺伝子のオン・オフ制御はこの抜粋記事ではとうてい説明しきれないものであり，全容が解明されたわけではない。遺伝子の塩基配列だけでなく，遺伝子以外のDNAや塩基配列以外の化学修飾も複雑に関与していることが判明，活発な研究が続いている。

なお，共著者のプタシュネ（Mark Ptashne，1940年～）は米国の生化学者で，1997年のアルバート・ラスカー基礎医学研究賞を受賞している。

初出：SCIENTIFIC AMERICAN June 1970
抜粋掲載：SCIENTIFIC AMERICAN July 2013, 日経サイエンス 2013年11月号

How are genes controlled? All cells must be able to turn their genes on and off. For example, a bacterial cell may need different enzymes in order to digest a new food offered by a new environment. As a simple virus goes through its life cycle its genes function sequentially, directing a series of timed events. As more complex organisms develop from the egg, their cells switch thousands of different genes on and off, and the switching continues throughout the organism's life cycle. This switching requires the action of many specific controls. During the past 10 years one mechanism of such control has been elucidated in molecular terms: the control of specific genes by molecules called repressors. Detailed understanding of control by repressors has come primarily through genetic and biochemical experiments with the bacterium *Escherichia coli* and certain viruses that infect it.

The repressor binds, or attaches, directly to the DNA molecule at the beginning of the set of genes it controls, at a site called the operator, preventing the RNA polymerase from transcribing the gene into RNA and thus turning off the gene. Each set of independently regulated genes is controlled by a different repressor made by a different repressor gene.

The repressor determines when the gene turns on and off by functioning as an intermediate between the gene and an appropriate signal. Such a signal is often a small molecule that sticks to the repressor and alters or slightly distorts its shape. In some cases this change in shape renders the repressor inactive, that is, no longer able to bind to the operator, and so the gene is no longer repressed; the gene turns on when the small molecule, which here is called an inducer, is present. In other cases

Vocabulary

gene 遺伝子
enzyme 酵素
digest 消化する
life cycle 生活環
develop 発生する, 発達する
elucidate 解明する
repressor 抑制因子, リプレッサー
▶ Technical Terms
Escherichia coli 大腸菌
RNA polymerase RNA ポリメラーゼ
▶ Technical Terms
transcribe 転写する
independently 独立に
intermediate 仲介者
distort 変形させる, 歪める
render ～を～にする
inducer インデューサー

Technical Terms

抑制因子, リプレッサー（**repressor**） 特定の遺伝子の発現を抑える働きをするタンパク質。
RNA ポリメラーゼ（**RNA polymerase**） DNA を鋳型として RNA 鎖を合成する反応を触媒する酵素。これは DNA の情報を RNA に転写する反応なので, 転写酵素とも呼ばれる。

the complex of the repressor and the small molecule is the active form; the repressor is only able to bind to the operator when the small molecule (here called a corepressor) is present.

Richard Burgess and Andrew Travers of Harvard University and Ekkehard Bautz and John J. Dunn of Rutgers University have shown that RNA polymerase, which initiates the synthesis of RNA chains at the promoters, contains an easily dissociated subunit that is required for proper initiation. This subunit, the sigma factor, endows the enzyme to which it is complexed with the ability to read the correct promoters. Travers has shown that the *E. coli* phage T4 produces a new sigma factor that binds to the bacterial polymerase and enables it to read phage genes that the original enzyme-sigma complex cannot read. This change explains part of the timing of events after infection with T4.

The first proteins made are synthesized under the direction of the bacterial sigma factor; among these proteins is a new sigma factor that directs the enzyme to read new promoters and make a new set of proteins. This control by changing sigma factors can regulate large blocks of genes. We imagine that in *E. coli* there are many classes of promoters and that each class is recognized by a different sigma factor, perhaps in conjunction with other large and small molecules.

Both the turning on and the turning off of specific genes depend ultimately on the same basic elements we have discussed here: the ability to recognize a specific sequence along the DNA molecule and to respond to molecular signals from the environment. The biochemical experiments with repressors demonstrate the first clear mechanism of gene control in molecular terms. Our detailed knowledge in this area has provided some tools with which to explore other mechanisms.

Vocabulary

corepressor コリプレッサー

initiate 始める，開始させる
promoter プロモーター
dissociate 解離する
sigma factor シグマ因子
endow A with B　AにBを付与する
phage ファージ，バクテリオファージ。細菌に感染するウイルス

infection 感染

in conjunction with　～とともに，～と一緒に

遺伝子はどのように制御されているのだろうか？　細胞はみな，その遺伝子をオン・オフできなければならない。例えば細菌の細胞なら，新しい環境に入って以前とは異なる食べ物が提供されると，これを消化するために異なる酵素が必要になるだろう。単純なウイルスも，その生活環の各段階で遺伝子を順に機能させ，しかるべき時にしかるべき事柄を引き起こすように指令している。より複雑な生物が受精卵から発生する際には何千もの異なる遺伝子がオン・オフされ，こうした切り替えはその生物の一生を通じて続く。この遺伝子スイッチングには多くの特定の制御が必要だ。過去10年で，そうした制御機構の1つが分子レベルで解明された。「リプレッサー（抑制因子）」という分子が特定の遺伝子を制御している。この制御機構の解明は主に，大腸菌とそれに感染するある種のウイルスを用いた遺伝学的・生化学的な実験によるものだ。

リプレッサーは，その制御対象となる遺伝子の始まりにあたるDNA領域に直接結びつく。この結合部位は「オペレーター」と呼ばれ，リプレッサーがそこに結合するとRNAポリメラーゼが遺伝子をRNAに転写するのが阻害され，したがってこの遺伝子はオフになる。制御対象となる遺伝子群はリプレッサーごとに異なり，それぞれのリプレッサーは異なるリプレッサー遺伝子によって作られている。

リプレッサーは遺伝子と適切なシグナルの間で仲介者として機能することによって，その遺伝子をいつオン・オフするかを決めている。そうしたシグナルは小さな分子であることが多く，これがリプレッサーにくっついて，その形をわずかに歪める。ある場合には，この形状変化がリプレッサーを不活性にし，つまりオペレーターに結合できなくして，この結果，当該の遺伝子は抑制されなくなる。つまりこの小分子が存在すると遺伝子はオンとなる。この場合の小分子は「インデューサー」と呼ばれる。その他の場合には，リプレッサーと小分子の複合体が積極的に遺伝子抑制に働く。この場合の小分子は「コリプレッサー」と呼ばれ，リプレッサーはこの小分子が存在するときにだけ，オペレーターに結合できる。

ハーバード大学のバージェス（Richard Burgess）とトラバース（Andrew Travers），およびラトガーズ大学のバウツ（Ekkehard Bautz）とダン（John J.

Dunn）は，DNAのプロモーターに結合してRNA鎖の合成を開始するRNAポリメラーゼが容易に解離するサブユニットを含んでおり，このサブユニットが適切な合成開始に必要であることを示した。このサブユニットは「シグマ因子」と呼ばれ，シグマ因子を含むRNAポリメラーゼだけが正しいプロモーターを読むことができる。トラバースは大腸菌に感染するT4ファージが大腸菌のポリメラーゼに結合する新たなシグマ因子を作り出し，大腸菌本来のポリメラーゼ・シグマ因子では読めなかったファージの遺伝子を読めるようにしていることを示した。この変化によって，T4感染後に起こる出来事の順序に部分的説明がつく。

最初のころにできるタンパク質は大腸菌のシグマ因子の指令の下に作られるが，そのなかには新たなシグマ因子が含まれ，これがポリメラーゼに新たなプロモーターを読むよう指令して，一連の新たなタンパク質が作られるのだ。シグマ因子の変化によるこの制御は，多数の遺伝子をひとまとめにして調節できる。私たちは，大腸菌には多くの種類のプロモーターがあり，それぞれが異なるシグマ因子によって認識されるのだろうと考えている。おそらくはシグマ因子とは別の大小の分子とともに。

特定の遺伝子のオンもオフもつまるところ，ここで述べてきた同じ基本要素によっている。DNA分子鎖に沿って特定の配列を認識する能力と，周囲の環境からの分子シグナルに反応する能力だ。リプレッサーについての一連の生化学実験は，遺伝子制御のメカニズムを分子レベルで初めて明確に示した。この分野の詳細な知識は他の生化学メカニズムを探るいくつかの手段をもたらした。

RNA as an Enzyme

酵素機能をもつRNA（1986年掲載）

T. R. チェック（1989年受賞）

トーマス・ロバート・チェック（Thomas Robert Cech，1947年～）は米国の分子生物学者・生化学者。1989年，「RNAの触媒機能の発見」によって，米国の分子生物学者シドニー・アルトマン（Sidney Altman, 1939年～）とともにノーベル化学賞を受賞した。

DNAの遺伝情報がRNAへ変換される「転写」の仕組みを研究するなかで，原生動物のテトラヒメナから抽出したRNAが自分自身を切断する反応を引き起こすことを発見した。つまりRNAは遺伝情報を記録しているだけでなく，触媒の機能を果たして細胞内での反応に関与している。チェックはこのRNAを「リボザイム」と名づけた。

以下の記事はノーベル賞受賞の3年前に執筆された。授賞業績であるリボザイムの発見について解説し，リボザイムが持つ意味を考察している。「リボソーム（タンパク質を合成している細胞小器官）ではタンパク質ではなくRNAが触媒として作用している可能性も考えられる」と述べているが，これはその後，実際に確認された。

「DNAやRNAは遺伝情報の担い手，タンパク質は生化学反応の主役」という，わかりやすいが単純な図式が崩れたことで，生命科学の研究はますます面白くなったといえるだろう。

初出：SCIENTIFIC AMERICAN November 1986, サイエンス 1987年1月号
抜粋掲載：SCIENTIFIC AMERICAN July 2013, 日経サイエンス 2013年11月号

In a living cell the nucleic acids DNA and RNA contain the information needed for metabolism and reproduction. Proteins, on the other hand, are functional molecules: acting as enzymes, they catalyze each of the thousands of chemical reactions on which cellular metabolism is based. Until recently it was generally accepted that the categories are exclusive. Indeed, the division of labor in the cell between informational and catalytic molecules was a deeply held principle of biochemistry. Within the past few years, however, that neat scheme has been overturned by the discovery that RNA can act as an enzyme.

The first example of RNA catalysis was discovered in 1981 and 1982 while my colleagues and I were studying an RNA from the protozoan *Tetrahymena thermophila*. Much to our surprise, we found that this RNA can catalyze the cutting and splicing that leads to the removal of part of its own length. If one could overlook the fact that it was not a protein, the *Tetrahymena* RNA came close to fulfilling the definition of an enzyme.

What does the startling finding of RNA enzymes imply? The first implication is that one can no longer assume a protein lies behind every catalytic activity of the cell. It now appears that several of the operations that tailor an RNA molecule into its final form are at least in part catalyzed by RNA. Moreover, the ribosome (the organelle on which proteins are assembled) includes several molecules of RNA, along with a variety of proteins. It may be that the RNA of the ribosome—rather than its protein—is the catalyst of protein synthesis, one of the

Vocabulary

nucleic acid 核酸
metabolism 代謝
reproduction 増殖
protein タンパク質
enzyme 酵素
catalyze 触媒する

exclusive 排他的，重ならない

neat きちんとした

protozan 原生動物の
Tetrahymena thermophila テトラヒメナ

overlook 見逃す，大目に見る

startling びっくりさせる，驚くべき
imply 意味する，含意する
lie behind （〜の原因として）背後にある
tailor こしらえる，作る
ribosome リボソーム ▶ Technical Terms
organelle 細胞小器官

Technical Terms

リボソーム（**ribosome**） 細胞内でタンパク質合成の場となっている構造体。メッセンジャーRNAの情報がここで読み取られ，それに従って適切なアミノ酸が運び込まれてつながっていく。リボソーム自体はRNAとタンパク質の複合体。

most fundamental biological activities. RNA catalysis also has evolutionary implications. Since nucleic acids and proteins are interdependent, it has often been argued that they must have evolved together. The finding that RNA can be a catalyst as well as an informational molecule suggests that when life originated, RNA may have functioned without DNA or proteins.

Having wandered back into the prebiotic past, it is fun to peer into the future and speculate about where the next examples of RNA catalysis might be found. In all known examples the substrate for the RNA enzyme has been RNA: another part of the same molecule, a different RNA polymer or a single nucleotide. This is probably not accidental. RNA is well suited to interacting with other RNAs, but it is more difficult to envision RNA forming a good active site with other biologically significant molecules such as amino acids or fatty acids. Hence I expect that future examples of RNA catalysis will also entail RNA as the substrate.

Two possibilities come to mind. One involves the small nuclear ribonucleoprotein particles (snRNPs) required for many operations in the nucleus. The other possibility is the ribosome.

The conclusion that protein synthesis is catalyzed by RNA would be a final blow to the idea that all cellular

Vocabulary

interdependent 相互に依存している

prebiotic 生物以前の，生命が存在する以前の

substrate 基質
▶ Technical Terms

nucleotide ヌクレオチド
accidental 偶然の

active site 活性部位
amino acids アミノ酸
fatty acid 脂肪酸

entail 必要とする

snRNPs　核内低分子 RNA-タンパク質複合体粒子，スナープス
▶ Technical Terms

final blow 決定的な打撃

Technical Terms

基質（**substrate**）　酵素の作用を受けて反応を起こす物質を，その酵素に対する基質という。つまり，酵素が作用する相手物質のこと。

核内低分子 RNA-タンパク質複合体粒子，スナープス（**small nuclear ribonucleoprotein particles；snRNPs**）　細胞核のなかに見られる様々な小さな RNA 分子は核内低分子 RNA（**snRNA**）と総称され，一般にタンパク質と複合体を形成しているので，この複合体を核内低分子 RNA-タンパク質複合体と呼んでいる。果たしている機能はいろいろあるが，mRNA 前駆体のスプライシングが代表的。DNA から転写された RNA 鎖に作用して，タンパク質をコードしていない部分（イントロン）を切除して mRNA に仕上げる反応だ。つまり，snRNA は別の RNA に対する酵素として反応を触媒している。

function resides in proteins. Of course, it may not be so; the ribosome may be such an intimate aggregation of protein and nucleic acid that its catalytic activity cannot be assigned exclusively to either component. Yet whether or not the synthetic activity of the ribosome can be attributed to the ribosomal RNA, a fundamental change has taken place in biochemistry in the past five years. It has become evident that, in some instances at least, information-carrying capacity and catalytic activity inhere in the same molecule: RNA. The implications of this dual capacity are only beginning to be understood.

Vocabulary

intimate 密接に関連した
aggregation 集合体

attribute 帰属させる

inhere 性質などが固有に備わっている

DNAとRNAは生きている細胞の中で代謝と増殖に必要な情報を担っている。これに対しタンパク質は機能分子であり，細胞の代謝の基礎となる多数の化学反応をそれぞれ触媒する酵素として働いている。最近まで，この区分は重なっていないとされていた。実際，情報分子と機能分子という細胞内での役割分担は生化学の基本原理となっていた。しかし過去数年で，このきちんとした図式が，RNAが酵素として作用しうるという発見によって覆された。

RNA酵素の最初の例は，私たちが原生動物のテトラヒメナから抽出したRNAを調べるなかで1981年から1982年にかけて発見された。たいへん驚いたことに，このRNAは切断とスプライシングの反応を触媒して，自分自身の一部分を除去できることがわかったのだ。タンパク質ではないということに目をつぶれば，テトラヒメナのRNAは酵素としての条件をほぼ満たしている。

この驚くべきRNA酵素の発見は何を意味するだろうか。まず第1に，細胞中の触媒反応がすべてタンパク質酵素によるという既成概念はもはや成り立たない。あるRNA分子を最終的な形に仕上げる過程のいくつかの段階が，少なくとも部分的にはRNAによって触媒されていることが明らかになったのだ。さらに，リボソーム（タンパク質合成を行う細胞小器官）は様々なタンパク質とともに，いくつかのRNA分子で構成されている。したがって，リボソーム中ではタンパク質ではなくRNAが，最も基礎的な生物学的反応であるタンパク質合成の触媒として作用している可能性も考えられる。RNA酵素は進化の問題にとっても意味がある。核酸とタンパク質は互いに依存し合っているので，両者はともに進化してきたに違いないとしばしば主張されてきた。しかしRNAが情報分子としてだけでなく酵素としても機能しうるという発見は，生命が誕生したときにはDNAやタンパク質なしにRNAが機能していた可能性をうかがわせる。

生物以前の太古をこうして逍遥したところで未来に目を転じ，RNA酵素の新たな例が見つかる可能性について思いをめぐらせるのも面白い。これまで知られている例では，RNA酵素の基質はすべてRNAであり，同じ分子の別の部分であるか，別のRNAポリマーまたは単一のヌクレオチドである。これはおそらく偶然ではないだろう。RNAは他のRNAと反応するにふさわしいが，アミノ酸や脂肪酸など他の生物学的に重要な分子と適切な活性部位を作るとは考えにくい。したがって私は，今後見つかるRNA酵素の例も，その基質はRNAだろうと考えている。

2つの可能性が思い浮かぶ。1つは細胞核内での様々な反応に必要とされている核内低分子RNA-タンパク質複合体粒子（snRNP，スナープス）であり，もう1つはリボソームである。

基本的な生合成反応であるタンパク質合成がRNAの触媒作用によっているという結論は，細胞のすべての機能がタンパク質に依存しているという考え方に決定的な打撃となるだろう。もちろん，そうではないかもしれない。リボソームはタンパク質とRNAが密接にからみ合った複合体であり，触媒機能がどちらの構成成分に存在するのか明確に区別できない可能性もある。しかし，リボソームの合成活性がリボーソームRNAに帰属するものであろうとなかろうと，過去5年間に生化学分野で根本的な変革が起こったのは確かだ。少なくともいくつかの例で，情報を担う能力と触媒活性がRNAという同じ1つの分子に備わっていることが明らかになったのである。この二重の能力が持つ意味に関しては，今まさにわかり始めたばかりだ。

Plastics That Conduct Electricity

電気を通すプラスチック（1988年掲載）

R. B. カナー／ A. G. マクダイアミッド（2000年受賞）

アラン・グラハム・マクダイアミッド（Alan Graham MacDiarmid，1927～2007年）はニュージーランド出身の米国の化学者。2000年，「導電性高分子の発見と開発」によって，日本の化学者である白川英樹（1936年～）および米国の物理学者アラン・ジェイ・ヒーガー（Alan Jay Heeger，1936年～）とともにノーベル化学賞を受賞した。3人は1970年代後半，ペンシルベニア大学のマクダイアミッドの研究室でポリアセチレンに臭素やヨウ素を添加すると電気伝導度が劇的に向上することを発見した。電気を通すプラスチック，導電性高分子の誕生だ。

以下の記事はノーベル賞受賞の12年前，初の導電性高分子の合成から10年ほど後に書かれた。その時点ですでにバッテリーの電極などに実証されており，現在までに透明タッチパネルやプラスチック電池，リチウムイオン電池の電極などとして実用化している。

導電性高分子のインパクトが絶縁性プラスチックがもたらしたインパクトに匹敵するものになるかどうかは，抜粋記事の最後の段落にあるように，「時のみが語ってくれるだろう」。173ページの「巨大分子はいかに作られるか」と併せて読むと興味深い。

共著者のカナー（Richard B. Kaner）は米国の化学者で，執筆当時はカリフォルニア大学ロサンゼルス校の准教授。なお，サイエンス（日経サイエンスの前身）1988年4月号に掲載された初出記事の翻訳は当時筑波大学教授だった白川氏による。

初出：SCIENTIFIC AMERICAN February 1988, サイエンス 1988年4月号
抜粋掲載：SCIENTIFIC AMERICAN July 2013, 日経サイエンス 2013年11月号

To most people the title of this article would have seemed absurd 20 years ago, when conceptual prejudice had rigidly categorized plastics as insulators. The suggestion that a plastic could conduct as well as copper would have seemed even more ludicrous. Yet in the past few years these feats have been achieved through simple modifications of ordinary plastics. Called conducting polymers, the new materials combine the electrical properties of metals with the advantages of plastics that stirred such excitement in the 1930s and 1940s.

To make a polymer conduct electricity, small quantities of certain chemicals are incorporated into the polymer by a process called doping. The procedure for doping polymers is much simpler than the one used to dope classical semiconductors such as silicon.

Once the potential of polymers as conductors had been demonstrated, the idea took off. In 1977 the first conducting polymer was synthesized; in 1981 the first battery with polymer electrodes was demonstrated. Last summer conducting polymers matched the conductivity of copper, and a few months ago the first rechargeable polymer battery was put on the market.

Subsequent advances suggest that polymers may be made that conduct better than copper; better, indeed, than any other material at room temperature. They may even replace copper wires in circumstances where weight is a limiting factor, as in aircraft. Conducting polymers also have interesting optical, mechanical and chemical properties that, taken together with their ability to conduct, might make them effective in novel applications where copper would not do. For instance, thin polymer layers on windows could absorb sunlight, and the degree of tinting could be controlled by means of an applied electric potential.

Vocabulary

prejudice 先入観
insulator 絶縁体
conduct (電気を)伝導する
ludicrous ばかばかしい
feat 偉業, 離れ業

conducting polymer 導電性高分子

property 特性
stir かきたてる

incorporate 導入する, 組み込む
doping ドーピング

semiconductor 半導体

demonstrate 実証する
take off 急に勢いづく
synthesize 合成する

match 匹敵する

put on 出す, 登場させる

subsequent その後の

replace 置き換える
limiting factor 制約的要因

take together ひとまとめにして考える

tinting 着色, 色合い
electric potential 電圧(電気的ポテンシャル)

The human body is another "device" in which conducting polymers might someday play a part. Because they are inert and stable, some polymers have been considered for neural prostheses—artificial nerves. Polypyrrole in particular is thought to be nontoxic and can reliably deliver an appropriate electric charge. The dopant ion here might be heparin, a chemical that inhibits the clotting of blood and is known to function quite adequately as a dopant in polypyrrole. Alternatively, polymers could act as internal drug-delivery systems, planted inside the body and doped with molecules that double as drugs. The drug would be released when the polymer was transformed to its neutral state by a programmed application of an electric potential.

In many ways the status of conducting polymers in the mid-1980s is similar to that of conventional polymers 50 years ago. Although conventional polymers were synthesized and studied in laboratories around the world, they did not become technologically useful substances until they had been subjected to chemical modifications that took years to develop. Likewise, the chemical and physical properties of conducting polymers must be fine-tuned to each application if the products are to be economically successful. Regardless of the practical applications that might be found for conducting polymers, they will certainly challenge basic research in the years to come with new and unexpected phenomena. Only time will tell whether the impact of these novel plastic conductors will equal that of their insulating relatives.

Vocabulary

inert 化学的に不活性な
prostheses 人工装具
polypyrrole ポリピロール

heparin ヘパリン
clot (血液が)凝固する

drug-delivery system 薬剤送達システム
double 二役を務める

transforme 変化させる
application 印加, 加えること

conventional 在来の, 普通の, 伝統的な

modification 変更, 改変
chemical modification は化学的修飾

practical 実際的な
application 用途, 使い道

challenge 喚起する, 促し求める

プラスチックは絶縁体であるという先入観がまかり通っていた20年前なら，「電気を通すプラスチック」というこの記事のタイトルは馬鹿げて思えたに違いない。銅と同じくらい電気をよく通すプラスチックというと，さらに馬鹿げているとみられたはずだ。しかしここ数年で，ごく普通のプラスチックに簡単な改良を施すことによって，そうした離れ業が達成された。導電性高分子と呼ばれるこの新素材は，1930年代から40年代にかけて話題をさらったプラスチックの利点と金属の電気特性を兼ね備えている。

電気を通す高分子を作るには，ある種の薬品を少しだけ高分子に加える「ドーピング」という処理をする。高分子へのドーピングは，シリコンなど古典的な半導体で行われているドーピングよりもはるかに簡単だ。

ひとたび高分子が導電体になる可能性が実証されると，関連の成果が堰を切ったようにあふれ出てきた。1977年には最初の導電性高分子が合成され，1981年には高分子を電極とした最初のバッテリーが実証された。昨1987年の夏には導電性高分子の伝導度は銅に匹敵するまでになり，数カ月後には充電可能な初のプラスチック電池が市販された。

その後の進歩を見ると，いずれは銅よりも電気をよく通す高分子が合成されるかもしれない。室温で他のいかなる物質よりも電気をよく通す物質だ。航空機のように重量が厳しく制限される環境では，銅線が導電性高分子に置き換えられるかもしれない。導電性高分子はまた，興味深い光学的，機械的，化学的性質を持っているので，導電性という性質を組み合わせると，銅では不可能な新しい応用が開ける可能性がある。例えば窓に薄い高分子膜を張って太陽光を吸収させることができるだろうし，その色合いと濃さを高分子膜にかける電圧で制御できるだろう。

また，いつの日にか導電性高分子が役に立つ“デバイス”に人体がある。導電性高分子は無害で安定だから，ある種の導電性高分子は神経軸索の代わりになるかもしれない。つまり，人工神経になりうるだろう。とりわけポリピロールは無毒であり，高い信頼性をもって適切な電荷を輸送できる。この場合のドーパントイオンは，血液の凝固を防止する作用がありポリピロールのドーパントとして極めて適切なヘパリンが使われるだろう。あるいは，埋め込み型の薬剤送達システムとしても使えるだろう。薬としても働くドーパントを導電性高分子に加えて，これを人体に埋め込む。電圧を計画的に加えることによって高分子が中性の状態になると，薬が放出される。

1980年代半ばにおける導電性高分子の状況は，多くの点で50年前の普通の高分子の状況に似ている。当時，在来型の高分子は世界中の研究室で合成され実験されていたが，技術的に有用な材料になったのは化学的修飾が施されるようになってからで，その開発には何年もかかった。同様に導電性高分子も，経済的に成功する製品になるには，それぞれの用途に応じて化学的特性と物理的特性を微調整しなくてはならない。導電性高分子に今後どのような実際的な用途が見つかるにしろ，それがもたらす予想外の新現象についての基礎研究が促されるのは確実だ。そうした新しい導電性高分子のインパクトが絶縁性プラスチックがもたらしたインパクトに匹敵するものになるかどうかは，時のみが語ってくれるだろう。

Filming the Invisible in 4-D

極微の世界をとらえるナノムービー（2010年掲載）

A. H. ズウェイル（1999年受賞）

アハメッド・ハッサン・ズウェイル（Ahmed Hassan Zewail，1946年〜）はエジプト生まれの米国の化学者。「フェムト秒分光学を用いた化学反応の遷移状態の研究」によって1999年のノーベル化学賞を単独で受賞した。進行中の化学反応をフェムト秒（1フェムト秒は1000兆分の1秒）という非常に短い時間単位でとらえて，その化学反応がなぜ進むのか，反応の途中でどのような中間体ができているのかなどを探るアプローチだ。

以下の記事はノーベル賞受賞から10年ほど後に執筆された。分子のスチル写真だけでなく，それを組み合わせた“動画”を構成することで，物質の振る舞いの時間変化を分子レベルで観察する技術について述べている。

やや専門的になるため抜粋記事では詳しく述べられていないが，フェムト秒化学では対象の物質にレーザーを照射して化学反応を開始させ，今度は別のレーザーを当てて状態を調べるという操作を制御された時間間隔で繰り返すなど，高度なテクニックを使う。データ処理にコンピューターが欠かせないのはもちろんだ。X線による結晶構造解析（88ページ）が20世紀の科学に大きく貢献したように，新たな測定・可視化技術が科学のフロンティアを広げていくことだろう。

初出：SCIENTIFIC AMERICAN August 2010, 日経サイエンス2010年11月号
抜粋掲載：SCIENTIFIC AMERICAN July 2013, 日経サイエンス2013年11月号

The human eye is limited in its vision. We cannot see objects much thinner than a human hair (a fraction of a millimeter) or resolve motions quicker than a blink (a tenth of a second). Advances in optics and microscopy over the past millennium have, of course, let us peer far beyond the limits of the naked eye, to view exquisite images such as a micrograph of a virus or a stroboscopic photograph of a bullet at the millisecond it punched through a lightbulb. But if we were shown a movie depicting atoms jiggling around, until recently we could be reasonably sure we were looking at a cartoon, an artist's impression or a simulation of some sort.

In the past 10 years my research group at the California Institute of Technology has developed a new form of imaging, unveiling motions that occur at the size scale of atoms and over time intervals as short as a femtosecond (a million billionth of a second). Because the technique enables imaging in both space and time and is based on the venerable electron microscope, I dubbed it four-dimensional (4-D) electron microscopy. We have used it to visualize phenomena such as the motion of sheets of carbon atoms in graphite vibrating like a drum after being "struck" by a laser pulse, and the transformation of matter from one state to another. We have also imaged individual proteins and cells.

Although 4-D microscopy is a cutting-edge technique that relies on advanced lasers and concepts from quantum physics, many of its principles can be understood by considering how scientists developed stop-motion photography more than a century ago. In particular, in the 1890s, Étienne-Jules Marey, a professor at the Collège de France, studied fast motions by placing a rotating disk with slits in it between the moving object and a photographic plate or strip, producing a series of exposures similar to modern motion picture filming.

Vocabulary

resolve 分析する, 分離する, 解像する
microscopy 顕微鏡法
exquisite 精巧な
jiggle 小刻みに揺れる
time interval 時間間隔
femtosecond フェムト秒。フェムトは 1000 兆分の 1
venerable 由緒ある
visualize 可視化する
graphite グラファイト, 黒鉛
transformation 変化, 変換, 状態の遷移
cutting-edge 最先端の
quantum physics 量子物理学
photographic plate 写真乾板

Among other studies, Marey investigated how a falling cat rights itself so that it lands on its feet. With nothing but air to push on, how did cats instinctively perform this acrobatic feat without violating Newton's laws of motion? The fall and the flurry of legs took less than a second—too fast for the unaided eye to see precisely what happened. Marey's stop-motion snapshots provided the answer, which involves twisting the hindquarters and forequarters in opposite directions with legs extended and retracted.

If we wish to observe the behavior of a molecule instead of a feline, how fast must our stroboscopic flashes be? My group attacked this challenge by developing single-electron imaging, which built on our earlier work with ultrafast electron diffraction. Each probe pulse contains a single electron and thus provides only a single "speck of light" in the final movie. Yet thanks to each pulse's careful timing and another property known as the coherence of the pulse, the many specks add up to form a useful image of the object.

Single-electron imaging was the key to 4-D ultrafast electron microscopy (UEM). We could now make movies of molecules and materials as they responded to various situations, like so many startled cats twisting in the air.

My colleagues and I investigated how quickly a short length of protein would fold into one turn of a helix by heating the water in which the protein was immersed—a so-called ultrafast temperature jump. (Helices occur in innumerable proteins.) We found that short helices formed more than 1,000 times faster than researchers have thought—arising in hundreds of picoseconds to a few nanoseconds rather than the microseconds commonly believed. Knowing that such rapid folding occurs may lead to new understanding of biochemical processes, including

Vocabulary

right oneself 姿勢をまっすぐに回復する
land 着地する
flurry あわただしい動き
twist ひねる, ねじる
hindquarters 4本足動物の身体の後方部分
forequarters 同じく前方部分
extend 伸ばす
retract 引っ込める
feline 猫, ネコ科動物
single-electron imaging 単一電子画像法
diffraction 回折
coherence コヒーレンス
4-D UEM 4次元超高速電子顕微鏡
startle 驚かす
protein タンパク質
helix らせん
immerse 浸す
ultrafast temperature jump 超高速温度ジャンプ
innumerable 数え切れない
picosecond ピコ秒。ピコは1兆分の1
nanosecond ナノ秒。ナノは10億分の1

those involved in diseases.

Very recently, my Caltech group demonstrated two new techniques. In one, convergent-beam UEM, the electron pulse is focused and probes only a single nanoscopic site in a specimen. The other, near-field UEM, enables imaging of the evanescent electromagnetic waves ("plasmons") created in nanoscopic structures by an intense laser pulse—a phenomenon that underlies an exciting new technology known as plasmonics. This technique has produced images of bacterial cell membranes and protein vesicles with femtosecond- and nanometer-scale resolution.

The electron microscope is extraordinarily powerful and versatile. It can operate in three distinct domains: real-space images, diffraction patterns and energy spectra. It is used in applications ranging from materials and mineralogy to nanotechnology and biology, elucidating static structures in tremendous detail. By integrating the fourth dimension, we are turning still pictures into the movies needed to watch matter's behavior—from atoms to cells—unfolding in time.

Vocabulary

convergent-beam UEM 収束ビーム UEM
specimen 試料, 標本
near-field UEM 近接場 UEM
evanescent electromagnetic wave エバネッセント波 ▶ Technical Terms
plasmon プラズモン ▶ Technical Terms
intense 強い, 強度の大きな
plasmonics プラズモニクス
vesicle 小胞

versatile 用途の広い, 汎用性のある
energy spectrum エネルギースペクトル
mineralogy 鉱物学
elucidate 解明する

unfold 展開する, 進展する

Technical Terms

エバネッセント波(**evanescent electromagnetic wave**) 物質に電磁波を当てて反射させるとき, 条件によっては入射光の電磁場が物質の内部に侵入することがある。この電磁場をエバネッセント場といい, これがもとで発せられる電磁波がエバネッセント波。物質の表面近くのみにしみ出した光で, 近接場光ともいう。evanescent は「はかない」という意味。

プラズモン(**plasmon**) 金属中の自由電子集団の振動を量子(擬似的な粒子)として記述したもの。金属の超微粒子に光を当てると, 光のエネルギーが微粒子表面のプラズモンに変換される。こうした過程をうまく制御して光学やエレクトロニクスに役立てようという技術分野がプラズモニクスだ。

人の目に見えるものは限られている。毛髪（太さ0.1mmほど）よりはるかに細い物体は見えないし，まばたき（1/10秒）より速い動きを追うこともできない。しかし，何世紀もかけて光学と顕微鏡法が発展してきた結果，ウイルスの顕微鏡写真や電球を突き抜ける弾丸のストロボ写真（ミリ秒オーダー）など，肉眼の限界をはるかに超える世界の精巧な画像を見られるようになった。ただ，不規則に動き回る原子の動画を見せられたら，ただのアニメか想像図，CGだと考えただろう。無理もない。そんな撮影は最近まで不可能だったのだから。

カリフォルニア工科大学の私の研究グループは過去10年で新方式の画像化技術を開発し，原子スケールで1フェムト秒（フェムトは1000兆分の1）という短時間に起こる動きを解明してきた。この技術は空間（3次元）と時間の両方に関する画像化が可能であることに加えて電子顕微鏡に基づいているので，私は「4次元電子顕微鏡法」と命名した。私たちはこれを用いて，グラファイト（黒鉛）の炭素原子シートがレーザーパルスによって“打たれた”後に太鼓の膜のように振動する現象や，物質がある状態から別の状態へと遷移する様子などを画像化した。個々のタンパク質や細胞の画像化にも成功している。

4次元電子顕微鏡法は先進的なレーザーと量子物理学の概念に基づく最先端技術だが，その原理の多くは100年以上前のストップモーション写真の開発を考えることで理解できる。1890年代，コレージュ・ド・フランスの教授だったマレー（Étienne-Jules Marey）は，動く被写体と写真乾板の間にスリットつきの回転板を挿入して現代の動画撮影に似た連続シャッターを実現し，高速運動を研究した。

とりわけ，マレーは落下する猫が姿勢を立て直してきちんと着地する様子を調べた。押せるものといえば空気のみの状況で，猫はどのようにこの離れ業をニュートンの法則を犯さずに成功させるのか？　落下時間は1秒にも満たないので，何が起きているのか肉眼では正確にはわからない。マレーが撮影した一連のスナップショットから，落下中の猫が自らの足を伸ばしたり引っ込めたりして頭部側と尻側をそれぞれ逆向きにひねり，姿勢を立て直していることがわかった。

猫の落下ではなく分子の動きを観察したいとしたら，ストロボフラッシュをどれくらい速くしなければならないだろうか？　私のグループは，単一電子による画像化技術を開発することで，この難問に挑んだ。私たちの初期の研究成果である超高速電子線回折を基礎とする技術だ。各プローブパルスは1個の電子を含んでいるだけなので，最終的なムービーの中で1個の“光の点”を与えるにすぎない。だが，各パルスのタイミングを慎重に調整することに加え，パルスが持つコヒーレンスという性質のおかげで，そうした点を多数重ねると意味のある画像が構成される。

単一電子画像法が，4次元超高速電子顕微鏡（UEM）のカギとなった。今では，分子や物質が，驚いた猫たちが空中で体をひねっているかのように，様々な状況に反応している様子を撮影できるようになった。

私たちは短いタンパク質がどれほど速く折り畳んでらせん構造を形成するかを，タンパク質を浸した周囲の水を加熱することによって調べた。「超高速温度ジャンプ」と呼ばれる構造形成だ（多くのタンパク質がらせん構造をとる）。その結果，それまで考えられていたよりも1000倍以上も速く短いらせんが形成されることがわかった。広く信じられてきたマイクロ秒のオーダーではなく，数百ピコ秒から数ナノ秒で形成される。こうした高速の折り畳みを理解することは，疾病に伴う生化学プロセスなどの解明につながるだろう。

カリフォルニア工科大学の私のグループはごく最近，2つの新技術を実証した。1つは「収束ビームUEM」で，電子パルスを絞って試料のごく小さな部分のみを観察する。もう1つは「近接場UEM」で，強力なレーザーパルスを当てることで超微細構造に発生する短命な電磁波「エバネッセント波（プラズモン）」の画像化を可能にする。プラズモニクスと呼ばれる興味深い新技術の背景となっている現象だ。フェムト秒かつナノスケールの分解能で，細菌の細胞膜やタンパク質小胞の画像が得られている。

電子顕微鏡は非常に有用かつ多目的に利用でき，実空間画像，回折像，エネルギースペクトルという3つのモードで対象を観察できる。材料科学や鉱物学，ナノテクノロジー，生物学まで幅広く利用され，静的な構造を詳細に明らかにしてきた。そして私たちは4番目の次元を組み入れて静止画を動画に発展させた。原子から細胞まで，物質の振る舞いの時間変化を観察可能になったのだ。

日経サイエンスで鍛える科学英語
ノーベル賞科学者編

2015年12月20日　1版1刷

編者　日経サイエンス編集部

発行者　竹内 雅人

発行所　日経サイエンス社
http://www.nikkei-science.com/

発売　日本経済新聞出版社
東京都千代田区大手町1-3-7　〒100-8066
電話 03-3270-0251(代)

印刷・製本 大日本印刷
ISBN978-4-532-52070-0